ELECTRICAL ENGINEERING DEVELOPMENTS

MAXIMUM POWER POINT TRACKING

BACKGROUND, IMPLEMENTATION AND CLASSIFICATION

ELECTRICAL ENGINEERING DEVELOPMENTS

Additional books and e-books in this series can be found on Nova's website under the Series tab.

ELECTRICAL ENGINEERING DEVELOPMENTS

MAXIMUM POWER POINT TRACKING

BACKGROUND, IMPLEMENTATION AND CLASSIFICATION

MAURICE HÉBERT
EDITOR

Copyright © 2020 by Nova Science Publishers, Inc.

All rights reserved. No part of this book may be reproduced, stored in a retrieval system or transmitted in any form or by any means: electronic, electrostatic, magnetic, tape, mechanical photocopying, recording or otherwise without the written permission of the Publisher.

We have partnered with Copyright Clearance Center to make it easy for you to obtain permissions to reuse content from this publication. Simply navigate to this publication's page on Nova's website and locate the "Get Permission" button below the title description. This button is linked directly to the title's permission page on copyright.com. Alternatively, you can visit copyright.com and search by title, ISBN, or ISSN.

For further questions about using the service on copyright.com, please contact:
Copyright Clearance Center
Phone: +1-(978) 750-8400   Fax: +1-(978) 750-4470   E-mail: info@copyright.com.

NOTICE TO THE READER

The Publisher has taken reasonable care in the preparation of this book, but makes no expressed or implied warranty of any kind and assumes no responsibility for any errors or omissions. No liability is assumed for incidental or consequential damages in connection with or arising out of information contained in this book. The Publisher shall not be liable for any special, consequential, or exemplary damages resulting, in whole or in part, from the readers' use of, or reliance upon, this material. Any parts of this book based on government reports are so indicated and copyright is claimed for those parts to the extent applicable to compilations of such works.

Independent verification should be sought for any data, advice or recommendations contained in this book. In addition, no responsibility is assumed by the Publisher for any injury and/or damage to persons or property arising from any methods, products, instructions, ideas or otherwise contained in this publication.

This publication is designed to provide accurate and authoritative information with regard to the subject matter covered herein. It is sold with the clear understanding that the Publisher is not engaged in rendering legal or any other professional services. If legal or any other expert assistance is required, the services of a competent person should be sought. FROM A DECLARATION OF PARTICIPANTS JOINTLY ADOPTED BY A COMMITTEE OF THE AMERICAN BAR ASSOCIATION AND A COMMITTEE OF PUBLISHERS.

Additional color graphics may be available in the e-book version of this book.

Library of Congress Cataloging-in-Publication Data

ISBN: 978-1-53618-164-7

Published by Nova Science Publishers, Inc. † New York

CONTENTS

Preface		**vii**
Chapter 1	Maximum Power Point Tracking in Solar PV Systems: A State of the Art *Deepak Verma, Nikhil Kumar and Savita Nema*	**1**
Chapter 2	MPPT Charge Controller for Battery Connected Photovoltaic Power Conditioning Unit *Joydip Jana, Hiranmay Samanta, Konika Das Bhattacharya and Hiranmay Saha*	**65**
Chapter 3	Comparison between SiC- and Si-Based Inverters Equipped with Maximum Power Point Tracking Charge Controller for Photovoltaic Power Generation Systems *Takeo Oku, Yuji Ando, Taisuke Matsumoto and Masashi Yasuda*	**109**
Chapter 4	Maximum Power Point Tracking: A Review of the Considerations for Large Scale and Small Scale Photovoltaic Installations *Sarah Lyden*	**137**

Chapter 5	Image Based Maximum Power Point Tracking in Wind Energy Conversion Systems *K. Sujatha, M. Malathi, N. P. G. Bhavani and V. Srividhya*	**163**
Index		**175**

PREFACE

Maximum Power Point Tracking: Background, Implementation and Classification presents state-of-art of existing conventional maximum power point techniques, along with shading mitigation techniques, and compares them on various parameters.

Photovoltaic systems include storage batteries when there is surplus power to provide electricity on demand. A suitable charge controller is needed for interfacing the solar photovoltaic module(s) with the battery bank. As such, attention has been made to attribute more features to the controller which will enhance the efficiency and controllability, and to monitor the health of the battery being charged.

The authors review the considerations for maximum power point tracking in large utility scale photovoltaic systems and small-scale residential photovoltaic systems. A set of characteristics is proposed and criteria is defined to evaluate the suitability of a technique.

In the penultimate study, power storage systems in ~100 W level are developed, which consist of direct current-alternating current converters, spherical Si solar cells, a maximum power point tracking controller, and lithium-ion batteries. Two types of inverters were used: SiC metal-oxide-semiconductor field-effect transistors (MOSFETs) and conventional Si MOSFETs.

In closing, the authors propose a simplified control stratagem to offer optimal power output power from a variable speed grid connected wind energy conversion system.

Chapter 1 - Under normal operating conditions, all solar photovoltaic (SPV) systems are operated at its Maximum Power Point (MPP). The MPP Tracker (MPPT) tracks all such optimal operating points that keeps shifting due to changing solar insolation, temperature or load. Under partial shading condition, the solar insolation is non-uniform and SPV array exhibits power characteristics with multiple peaks. The maxima of multiple peaks is termed as global peak and rest are local peaks. The multiple peaks on power characteristics of SPV array distract MPP Tracker and operating point may deviate from global peak to possibly settle at local maxima. The SPV array operation at local maxima delivers less power to load than available generation. The difference appears as system losses and must be minimized to have optimal conversion efficiency. The SPV array operation at its global peak thus guarantees transfer of maximum generated power to the load even under partial shading condition. The conventional MPPT technique fails to track global peak under partial shading or rapidly changing insolation. A special circuit arrangement or algorithm is therefore required to mitigate negative effect of partial shading. Multitude of shading mitigation techniques are available in literature that ensures global peak operation under partial shading condition, but each of them exhibits some vulnerabilities. This chapter *Maximum Power Point Tracking: Background, Implementation and Classification* therefore presents state-of-art of existing conventional MPPT techniques alongwith shading mitigation techniques and compares them on various parameters.

Chapter 2 - Photovoltaic (PV) systems include storage batteries when there is surplus power. This is used for providing electricity on demand. A suitable charge controller is needed for interfacing the solar PV module(s) with the battery bank. In this work, attention has been made to attribute more features to the controller which will enhance the efficiency and controllability and most importantly will monitor the health of the battery being charged, ensuring a long batterylife. Further, as the Maximum Power Point (MPP) of a solar PV array varies with solar irradiance and temperature, the developed controller has the facility to accurately track the MPP during static (fixed irradiance) and dynamic (rapidly changing irradiance) both weather conditions. This development focuses on the Battery charging from the PV modules at its dynamically changing MPP and at the same time follows a novel charging method which keeps checking on the condition of the battery health. An integrated software running from an embedded platform operating using the DSPIC microcontroller has been carried out. Both these aspects, of MPPT and efficient Battery charging were generally investigated separately earlier. The charge controller has been tested with Modules from 150W to 320W and has been found having minimum converter efficiency of 93.8%, and it can extract as much power as possible to charge the battery with an MPP tracking speed of 1 second and maximum MPP tracking efficiency of 99.9% with a fluctuation of 0.86% around the target MPP in static irradiation condition. And in dynamic irradiation conditions, the before mentioned performance parameters become 1.67 seconds, 98.5% and 0.50% respectively.

Chapter 3 - Power storage systems in ~100 W level were developed, which cosisted of direct current-alternating current converters, spherical Si solar cells, a maximum power point tracking controller, and lithium-ion batteries. Two types of inverters were used for the study: one is SiC metal-oxide-semiconductor field-effect transistors (MOSFETs) as switching devices while the other is conventional Si MOSFETs. In the present 100 W level inverters, the *on*-resistance was considered to have little influence on the efficiency. However, the SiC-based inverter exhibited an approximately 3% higher direct current-alternating current conversion efficiency than the Si-based inverter. Power loss analysis indicated that the higher efficiency resulted predominantly from lower switching and reverse recovery losses in the SiC MOSFETs compared with those in the Si MOSFETs.

Chapter 4 - A large variety of Maximum Power Point Tracking (MPPT) techniques have been proposed in the literature ranging from simple approximation techniques to those based on computationally complex algorithms. In general, the development of complex techniques is aligned with the goals of improving MPPT performance under non-uniform environmental conditions which could occur due to cell damage or shading between cells such as what would be typically seen in a residential environment. Photovoltaic (PV) systems are frequently deployed in residential and utility scale implementations and the requirements of these installations in terms of control can be quite distinct. This chapter will review the considerations of MPPT in large utility scale PV systems and small-scale residential PV systems identifying common requirements and differences in the appropriateness of MPPT techniques applied. A set of characteristics of large scale and small-scale PV installations will be proposed, and criteria defined to evaluate the suitability of a technique for application in a large scale or small-scale system. Common and emerging MPPT techniques will be evaluated against these criteria. The main contribution of the chapter is highlighting the considerations in selecting an appropriate MPPT technique based on the scale of the PV system.

Chapter 5 - This manuscript deals with a simplified control stratagem to offer optimal power output power from a variable speed grid connected Wind Energy Conversion System (WECS). A permanent magnet synchronous generator (PMSG) with variable speed turbine is coupled to the gear box, a bridge wave rectifier with blocking diode, a dc-to-dc converter with current controlled voltage source inverter which is of boost type. Output power can be maximized using Image based Maximum Power Point Tracking (IMPPT) algorithm from the wind turbine operating in the range of cut-in to rated wind velocity which is sensed using Discrete Fourier Transform (DFT). The IMPPT algorithm, DC–DC and DC–AC converters with Fuzzy Logic Control (FLC) are simulated using MATLAB software. The obtained simulation results show that the identification of potential windmills is very important with respect to grid integration.

In: Maximum Power Point Tracking
Editor: Maurice Hébert

ISBN: 978-1-53618-164-7
© 2020 Nova Science Publishers, Inc.

Chapter 1

MAXIMUM POWER POINT TRACKING IN SOLAR PV SYSTEMS: A STATE OF THE ART

Deepak Verma[1],*, PhD, Nikhil Kumar[2] and Savita Nema[2], PhD

[1]Department of Electrical & Electronics Engineering,
Birla Institute of Technology Mesra, Jaipur Campus, Jaipur, RJ, India
[2]Department of Electrical Engineering, Maulana Azad National Institute of Technology, Bhopal, MP, India

ABSTRACT

Under normal operating conditions, all solar photovoltaic (SPV) systems are operated at its Maximum Power Point (MPP). The MPP Tracker (MPPT) tracks all such optimal operating points that keeps shifting due to changing solar insolation, temperature or load. Under partial shading condition, the solar insolation is non-uniform and SPV

---

* Corresponding Author's E mail: deepakverma16@gmail.com

array exhibits power characteristics with multiple peaks. The maxima of multiple peaks is termed as global peak and rest are local peaks. The multiple peaks on power characteristics of SPV array distract MPP Tracker and operating point may deviate from global peak to possibly settle at local maxima. The SPV array operation at local maxima delivers less power to load than available generation. The difference appears as system losses and must be minimized to have optimal conversion efficiency. The SPV array operation at its global peak thus guarantees transfer of maximum generated power to the load even under partial shading condition. The conventional MPPT technique fails to track global peak under partial shading or rapidly changing insolation. A special circuit arrangement or algorithm is therefore required to mitigate negative effect of partial shading. Multitude of shading mitigation techniques are available in literature that ensures global peak operation under partial shading condition, but each of them exhibits some vulnerabilities. This chapter *Maximum Power Point Tracking: Background, Implementation and Classification* therefore presents state-of-art of existing conventional MPPT techniques alongwith shading mitigation techniques and compares them on various parameters.

Keywords: SPV array, MPPT, partial shading, shading mitigation

1. BACKGROUND

Electricity demand is growing with highest rate of all the energy consumed worldwide. Thus the mankind is facing a massive challenge of never ending increase in energy demand as a result of overall socio economic growth [1-2]. The declining fossil fuel resources and incredible rate of its consumption to battle the prevailing industrial revolution diverges us on the peak of consumption of fossil fuel. Incompatibility of conventional sources to fulfill this bottomless valley of energy requirements, energy security, and especially the sky-rocketing hike of fossil fuels prices gives a work force direction to invent compatible option [3-5]. Despite of these unprovoked concerns, the global warming as an unavoidable outcome of carbon emissions by the conventional energy sources proves to be a momentous driver for renewable energy sources deployment. Ubiquitous accessibility of renewable energies like solar and wind offers a striking solution to comply all these requirements. Continuous efforts of researchers have shown an increased efficiency in both the conversion and transport of these energy sources. Thus they arise as an attractive alternative option to conventional solution [6-7]. It is a trend which is almost certain to evolve in upcoming power generation. All-pervading and copious availability of solar energy has an outstanding potential to make a significant contribution to the world's energy needs. Two ways to extract the solar energy are solar thermal plants and solar cells i.e., photovoltaic cells. In prevailing renewable energy projects the photovoltaic (PV) cell is on the leading edge as the promising future energy technology option [8]. The direct conversion of solar radiation to electrical energy by PV cells has a number of significant advantages. However, its proficient extraction demands accomplishment of some significant challenges such as energy fluctuation, huge investment low energy conversion efficiency of module, and energy cost [9-10]. Reducing energy cost of PV system is a big issue since maintenance requirement is very low and the only real cost savings to be made is in efficiency enhancement. Recent literature reveals that research efforts target to enhance the power output of the module in terms of MPPT.

1.1. SPV Power Generation

A SPV power generation system consists of numerous components like cells, electrical connections and power conditioning unit to regulating and/or changing the electrical outputs. These systems are rated in peak watts (W_p) which is an amount of electrical power that a SPV system can deliver at standard test conditions (STC).

Photovoltaic comprises the technology to convert sunlight directly into electricity. The term *photo* means light and *voltaic* electricity. A PV cell, also known as solar cell is a semiconductor device that generates electricity when direct or indirect solar radiation falls on it.

The solar cell is basically a p-n junction made-up of a thin layer of semiconductor that generates electrical energy electricity when solar radiation or insolation falls on it. The p-n junction absorbs photon energy from insolation and converts it into electrical energy. Solar cells are commonly made-up of silicon, and when exposed to solar radiations absorbs photon energy thus breaking covalent bond and causing an electron to jump from lower energy level to higher energy level. The liberated electron becomes free in conduction band leaving behind a hole in valence band. The energy of photons in effect generates new electron-hole pairs (EHPs), if it carries energy greater than band gap of the material. The newly generated carriers (*electrons and holes*) try to get separated in the corresponding regions, if get separated the *N-type* semiconductor now contains increased number of electron, and *P-type* semiconductor contains increased number of holes, creating electric field or voltage that can drive current through electrical load. Some of the carriers which are not separated or recombine before separation, does not contribute to the current from the cell rather causes a loss of energy called as recombination loss. This phenomenon of generation of voltage by means of photon energy is called photovoltaic effect [11].

1.2. Solar Cell/Module/Array

A single PV cell generates DC potential in the range of 0.5 to 1 volts with small power output less than *2 W* at STC. To supply required power to load higher than *2 W*, a number of solar cells are connected in series or parallel and the combination is called solar module. A module which is glass encapsulation and aluminium framed consists of several solar cells connected electrically in series with a silver strip (*fingers*) between solar cells. The glass provides protection to the semiconductor material against the environmental factors and outer aluminium frame supports mounting thus giving mechanical strength to it.

The voltage of PV module is standardized for different applications. In fact, one typical PV module contains 32, 36 or 96 cells wired in series. Such series connected modules form a string and each such string generates sufficiently high voltage to meet rated voltage specification of

PV system. The rated power specification may then be achieved by enhancing current by connecting these strings in parallel and such series parallel arrangement of PV modules forms PV Array. The different arrangement of PV modules is shown in Figure 1.1.

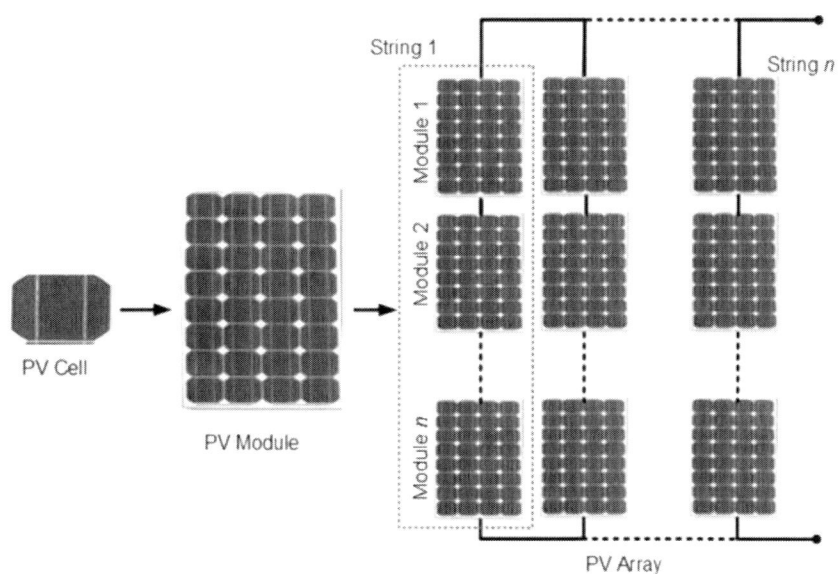

Figure 1.1. Formation of PV module and array.

1.3. Classification of Materials Used in Manufacturing

Silicon (Si) is most commonly occurring element after oxygen on the earth and more than 90% production of SPV modules are produced by Si wafers. Si has been workaholic for SPV industry, the first generation technologies of SPV system is based on Si wafers whereas the technologies based on thin film can be referred as second generation technologies. PV cells are classified based on the type of materials used in manufacturing them [11]. PV cells are basically classified into following categories.

1.3.1. Si Wafer Based Solar Cell Technologies

Wafer based silicon crystalline solar cells are blue cells which look like deep blue glass and have relatively high efficiency. These crystalline cells are most efficient in converting sunlight to electrical energy. However, the main disadvantage with this technology is the resulting high SPV cell price. The technology again is classified into two types as:

a) Monocrystalline
b) Polycrystalline

1.3.2. Thin Film Technologies

The thin film is semiconductor material of non-crystalline form. In thin film technology a thin layer of silicon is deposited on a base material such as metal or glass to create solar panel. The amorphous silicon cell was first solar cell based on thin film technology. The manufacturing process is simpler, easier and cheaper than in the crystalline cells. The producing method is less complicated, easier and cheaper than the crystalline cells. Although, the foremost common type of thin-film PV material is Si (silicon), nevertheless alternative materials such as CIGS (Copper Indium/Galliumdiselenide), CdTe (Cadmium Telluride), CIS (Copper Indium Selenide), dye-sensitized cells and organic solar cells may be used. The disadvantage of these cells is lower efficiency, around 6-8%, however have distinguished advantage of being versatile but costlier [11].

1.4. Balance of SPV Systems

An SPV system required several components other than PV generation unit or PV array, these components are referred as Balance of System (BoS) [11] that includes batteries, charge controllers, DC-DC converters, maximum power point tracker, DC-AC converters or inverters, peripheries/mountings and different control and protection circuits [12-15].

1.4.1. Battery

Batteries are one amongst the foremost sensitive and expensive part of an SPV system mainly used as backup in standalone (SA) or OFF-grid SPV system to fulfil the load demand when sun is not present. Relying upon the size and type of solar photovoltaic system, battery subsystem costs could vary from 10% to 50% of total system cost. The foremost drawback with battery is their inability to live up to expectations of user due to poor functioning of battery chargers and associated maintenance. This include overcharging, incomplete charging and prolonged operation at a low state of charge that result into enhanced running cost due to replacement of batteries before their expected life time.

1.4.2. Charge Controller

In SPV system the input charging voltage is from PV array and implicitly larger than battery voltage. A charge controller is employed to take care of proper charging voltage at terminals of battery. The charge controller additionally does not permit battery to discharge below a pre-set level under loading. This pre-set level is the one as specified by manufacturer. The charge controller is a device that regulates voltage from PV array and keeps battery among safe zone by loading it within allowed range of charging/discharging voltage. Usually PWM or constant voltage charge controllers are used. The charge controller is put in between PV array and battery (battery bank) to mechanically regulate charge on its electrodes. The range of circuits is employed in charge controller with additional options, and its suitability should be assessed in view of PV application [14].

1.4.3. DC-DC Converter

The changing nonlinear *I-V* (*current–voltage*) characteristics of PV module due to change of insolation and temperature causes terminal voltage of the module to change. This may deviate maximum power point due to which the available maximum power delivery to load can be differed. DC-DC converter is an essential part in tracking of maximum power, as the source can deliver maximum power only corresponding to

V_{MPP} (terminal voltage at which source deliver maximum power) or I_{MPP} (current of PV array corresponding to MPP). Shifting the terminal voltage (V_i) or current (I_i) is done by means of the DC-DC converters however the duty cycle is decided by the MPPT algorithm.

1.4.4. Maximum Power Point Tracker

The impact of environmental issues such as insolation and temperature as well as shading condition on solar cell characteristic enables the alteration in terminal voltage.

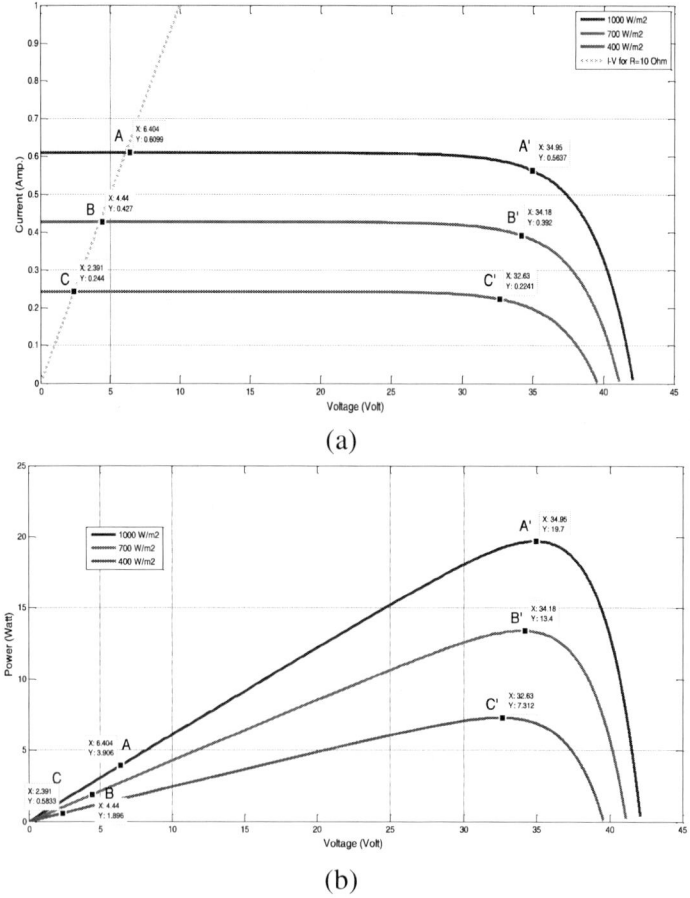

Figure 1.2. (a). Maximum power/operating points for different insolation on *I-V* curve, (b). Maximum power/operating points for different insolation on *P-V* curve.

The operating strategy of MPPT is explained by considering an example for tracking maximum power with change in insolation as shown in Figure 1.2(a). Figure presents the solar cell characteristic for three different insolation for a linear resistive load ($R_L = 10\Omega$), leading to different maximum power points which are *A', B' and C'* as shown in Figure 1.2(b). As the load is linear, the operating points and corresponding terminal voltages of the cell are *A, B and C* and it is clear from the Figure 1.2(a) and Figure 1.2(b) the power delivered by the solar cell with respect to point *A, B and C* is less than the available power.

The maximum power point tracking operating strategy evokes the concept of holding the terminal voltage at a value corresponding to the maximum power point i.e., *A', B' and C'* instead of operating point *A, B and C*. An electronic circuitry used to drag the operating point of solar cell to the maximum power point by means of DC-DC converter is known as maximum power point tracking [16].

1.4.5. DC-AC Converters or Inverters

The output of the PV array is direct current and cannot be fed directly into the grid or load; hence inverters are essentially used. The PV inverters are power electronic circuit that convert DC into AC power considering PV behaviour as power source [12].

1.5. Applications of SPV Systems

The application of SPV system is to use them as a supply system of an electrical energy for a particular load [16]. SPV systems need *BoS* to make it a reliable source of power.

SPV system can be classified into three categories:

a) Standalone (SA) PV system
b) Grid connected PV system
c) Hybrid PV system.

PV systems are becoming very popular as standalone systems, where grids are not available or uneconomical and difficult to install. Photovoltaic sources are used as standalone system in many applications. They have many advantages but they mainly suffer from problems like high installation cost and requirement of power conditioning devices (DC-DC or DC-AC converter) for load interface and lower PV panel efficiency. A typical stand-alone PV system consists of PV panels and battery.

When solar energy is not available battery is used to make electricity available to the load. The stand-alone PV system suffers from major drawbacks, the battery is costly and bulky and life is small. This can be avoided if the photovoltaic system is feeding the PV energy directly into the AC grid system. It offers the advantage of feeding the additional PV power into the grid thus reducing the amount of energy that has to be generated by conventional sources. When Sun is not available (*at night or on cloudy days*), the output of the PV system is insufficient, the grid provides energy from conventional sources. Grid connected PV system is advantageous and economical as it removes the battery, reducing cost and size of the whole system and also increases its reliability. The life of a PV cell is more than 20 years; the life of battery is hardly 3-4 years. The energy payback period (EPP) of the SPV system is about 5-6 years [17-18].

1.5.1. Standalone PV System

Standalone PV system can be simple or complex depending on the type of configuration. Consequently, the system configuration of standalone system can be divided into following categories [17-18]:

1.5.1.1. Uncontrolled Standalone System with DC Load

This is the simplest configuration of standalone PV system containing PV array and DC load as shown in Figure 1.3 (a). The disadvantage is that the generated PV power is not utilized optimally as well as no alternatives are there for energy storage to have night-time load operation.

1.5.1.2. Controlled Standalone System with DC Load

The disadvantage of uncontrolled standalone system is overcome with the use of an electronic controller as shown in Figure 1.3 (b). Electronic controller may be a charge controller or MPPT circuit to control the output of PV array for optimizing the power output of the PV array along with the DC-DC converter. This additional circuitry enhances the cost of the PV system but ensures the smooth and controlled operation of the load.

1.5.1.3. Controlled Standalone System with Battery and DC Load

Normally when MPPT is implemented, energy storage device is also used to absorb the extra energy generated by the PV. The other benefit of this configuration is to fulfil load requirement in night-time or when insolation is not available. Figure 1.3 (c) shows the controlled standalone system configuration with battery and DC load.

1.5.1.4. Controlled Standalone System with Battery, AC/DC Load

Most of the commonly used loads are of AC type so that this configuration contains one more component or BoS that is DC-AC converter or inverter which fulfil the load requirement of the AC appliances or load. In this configuration shown in Figure 1.3 (d) load requirement of both AC as well as DC type is fulfilled by the PV standalone system.

1.5.1.5. Controlled Hybrid System with AC/DC Load

In hybrid system other than PV system auxiliary source of power is there to avoid energy storage system this makes the configuration more reliable. In night time when solar radiation is not available, auxiliary source of power delivers the power to the load. The auxiliary source of power may be other renewable energy sources such as wind, biomass etc. or a diesel generator. To make this configuration more reliable the energy storage system may also be used.

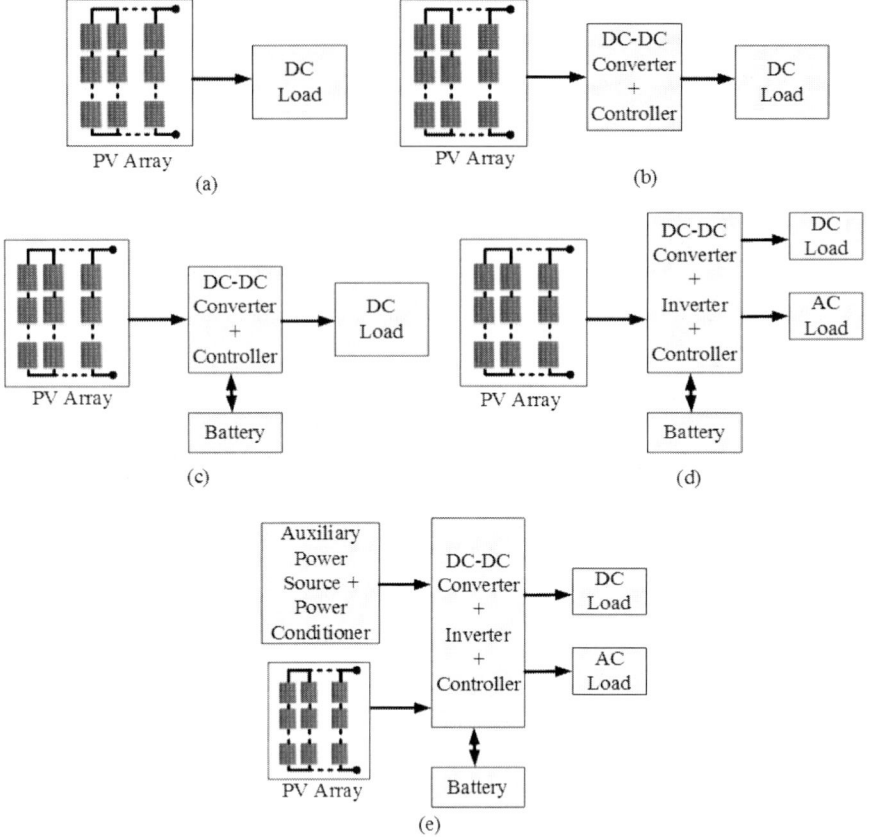

Figure 1.3. Block diagram representation of SA configurations: (a) Uncontrolled SA system with DC load (b) Controlled SA system with DC load (c) Controlled SA system with battery and DC load (d) Controlled SA system with battery, AC/DC load and (e) Controlled hybrid system with AC/DC load.

1.5.2. Grid connected PV System

Now a day's grid connected PV systems are gaining more popularity as they do not need any storage device and offers optimum use of PV generated energy. Application in distributed generation is also another major advantage of it. Controlled standalone system has the MPPT circuitry but in case of excess power production the excess power has to be stored in the battery bank which makes the system costly. Grid connected PV system offers the advantage of feeding excess power generated into the grid. The grid connected configuration may be single stage, two stage or multi stage conversion as shown in Figure 1.4. In single stage grid connected configuration only DC-AC converter is there, in this case optimum use of PV generated power is not done as the MPPT is not present. In two stage configuration there is intermediate stage between PV and DC-AC converter that is DC-DC converter stage along with the MPPT (*Stage 1 and stage 2*). Two stage grid connected configuration is more realistic as the optimum use of produced power is possible. In multistage configuration there is one more stage between DC-DC converter and DC-AC converter known as DC-DC conversion stage with galvanic isolation (*Stage 1, 2 and 3*). The multistage configuration has drawbacks such as higher component count, higher cost and large size.

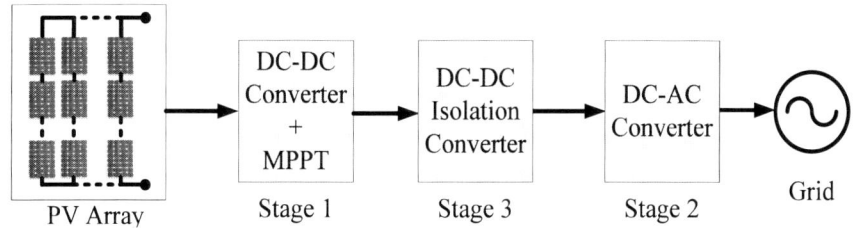

Figure 1.4. Block diagram representation of grid connected PV system.

1.5.3. Hybrid System

In hybrid system at least one alternative or auxiliary source is there in addition with the PV system as discussed in section 1.6.1.5. The applications of hybrid system are very large as they can be use as standalone configuration, grid connected system or in *micro grid* applications.

2. IMPLEMENTATION AND CLASSIFICATION OF MPPT TECHNIQUES

The fundamentals of MPPT problem and various techniques are discussed at length in this section. This section reviews the present state and trends in the evolution and development of maximum power point tracking techniques in solar PV systems. A detailed survey of the literature pertaining to the different aspects of MPPT and their comparison on the basis of 13 factors (Category, Dependency of PV array, Implementation methodology, Sensor required, Stages of energy conversion, Partial shading enabled, Grid interaction, Analog or Digital, Tracking efficiency, Tracking speed, Cost, Product available in market) is presented.

2.1. Maximum Power Point Tracking

The concept of maximum power point tracking is discussed in previous section, in this section some of the salient features of MPPT are discussed. The nonlinear I-V characteristic of a PV module is shown in Figure 2.1(a) and P-V characteristic shown in Figure 2.1(b) at standard test condition (STC) 1000 W/m^2 insolation, 25^{0}C temperature and 1.5 Air Mass. Where: I_{sc} is the short circuit current of the solar module, V_{oc} is the open circuit voltage of the solar module, I_m is the current at which module delivers maximum power and V_m is the terminal voltage of solar module at which module deliver maximum power. Some of the key aspects of the MPPT are:

- Maximum power can be transferred only at a specific terminal voltage for a given environmental condition known as maximum power point (MPP).
- MPP also changes with change in environmental conditions such as insolation temperature.
- MPP occurs normally at 70-80% of the open circuit voltage.
- Rate of change of power with respect to voltage (dP/dV) or current (dP/dI) becomes zero at MPP.
- Tracking of power is possible from both side as shown in Figure 2.1(b)

Maximum Power Point Tracking in Solar PV Systems 15

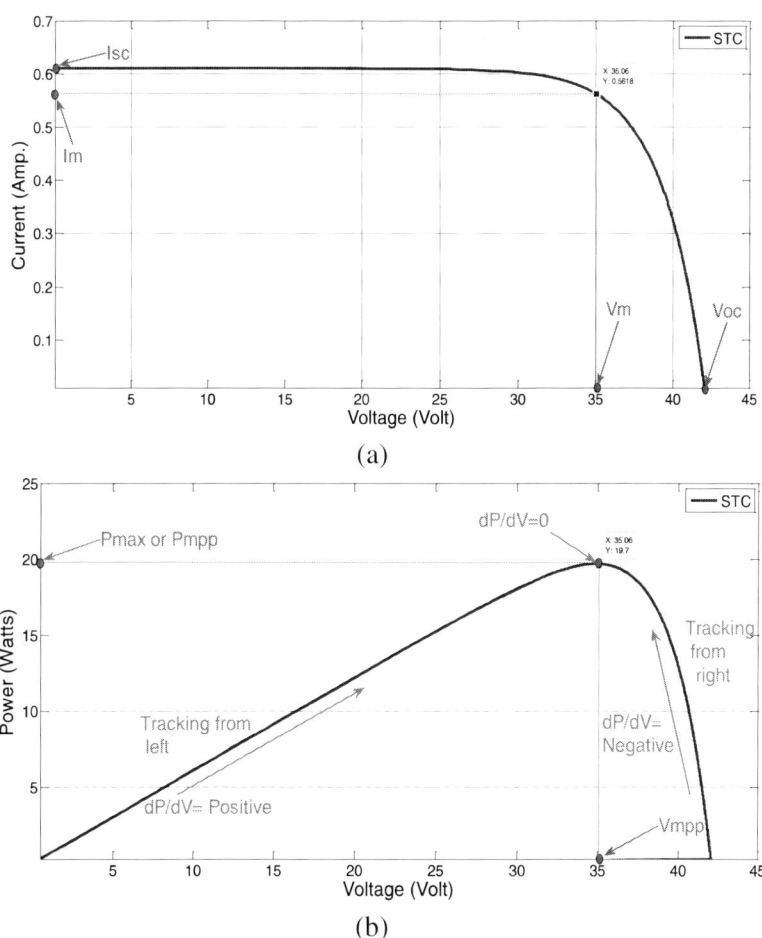

Figure 2.1. (a) I-V and (b) P-V characteristic of solar PV in context with MPPT.

2.2. Maximum Power Point Tracking Techniques

The literature review consists of survey of a number of papers from the various referred journals and conferences. The literature review gives adequate idea about the basics of the MPPT algorithm and operating strategy of different MPP tracking techniques.

The literature review consists of survey of a number of papers from the various referred journals and conferences. The literature review gives adequate idea about the basics of the MPPT algorithm and operating strategy of different MPP tracking techniques.

Leedy A W et al. [19] presents *constant voltage method*, Constant voltage method is based on the observation that the maximum power point occurs between 72-78% of the open circuit voltage V_{oc}, for the standard atmospheric condition. The solar PV module always operates at the constant voltage in this range. The duty ratio (D) of the DC-DC convertor ensures that the PV voltage is equal to:

$$V_{REF} = K_1 \times V_{oc} \qquad (2.1)$$

Where K_1 = 0.72 to 0.78

Figure 2.2 shows that after V_{OC} is sampled by a sampler, V_{REF} which is calculated by Eq. 2.1 is kept constant during one sampling period by hold circuit, now duty ratio D is adjusted to make $V_{PV} = V_{REF}$. For next sample again V_{OC} is sampled and the same procedure is repeated for each samples. Figure 2.3 depicts the flow chart of this method.

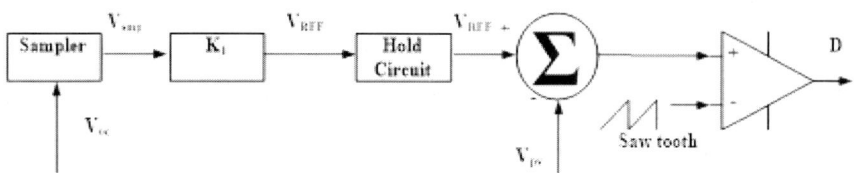

Figure 2.2. Maximum power point tracker using constant voltage method.

The method is simple, fast and easy to implement but shows limited accuracy, V_{oc} is required to be measured at regular interval and used only where lethargic temperature variation is observed.

Based on the same operating principle Salameh, Z. M et al. [20] presented a method named as *pilot cell method*, in this method a pilot cell is used to calculate the open circuit voltage instead of the whole PV array. After simple calculation array open circuit voltage i.e., V_{array} can be directly evaluated, which reduces the efforts of measuring V_{oc} at regular interval i.e., the problem of disconnection of PV from the load at every sample can be avoided. V_{REF} or voltage corresponding to the MPP can be calculated by Eq. 2.2:

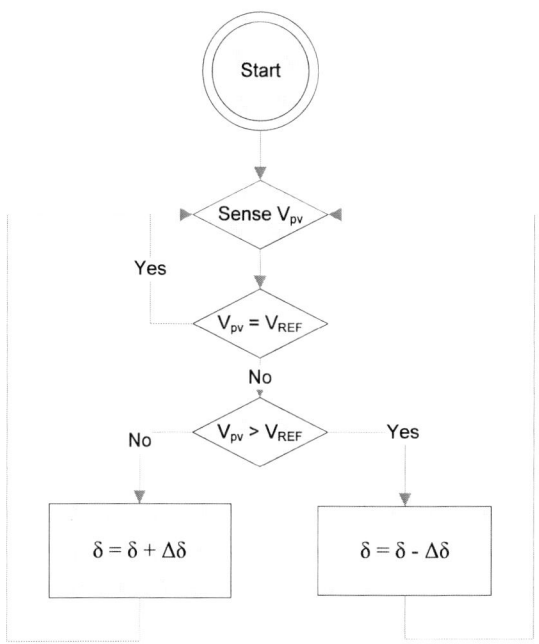

Figure 2.3. Flow chart of constant voltage method.

$$V_{REF} = K_3 \times V_{oc/pilot\ cell} \qquad (2.2)$$

Where $K_3 \approx$ constant < 1.

Alghuwainem, S M et al. [21] discusses *constant current method*. Constant current method is based on the same phenomenon of the constant voltage method. In the constant voltage method the PV array operates at the constant voltage and in this method PV array operates at the constant current. The maximum power point arrives between *78-92%* of the short circuit current I_{sc} thus the sensed parameter is short circuit current.

$$I_{REF} = K_2 \times I_{sc} \qquad (2.3)$$

Where $K_2 = 0.78$ to 0.92.

Leedy, A W et al. [22] proposed a modified *curve fitting method*, the characteristic of solar PV is non linear, which can be modeled mathematically by using curve fitting method. The nonlinear characteristic of a solar array can be approximated as:

$$P_{PV} = K_4 V_{PV}^3 + K_5 V_{PV}^2 + K_6 V_{PV} + K_7 \qquad (2.4)$$

At maximum power point, $dP_{PV}/dV_{PV} = 0$

$$V_M = \frac{-K_5 \sqrt{K_5^2 - 3K_4 K_6}}{3K_4} \qquad (2.5)$$

Where K_4, K_5, K_6, K_7, are constant and can be obtained by polyfit command in MATLAB.

Also it has been shown that P_{PV} is a function of array voltage and temperature. [22] gives a modified curve fitting method that predicts the P-V characteristic curve of a PV array with a fourth-order polynomial for varying temperatures.

$$\begin{aligned}P_{PV}(V_{PV}, T_{array}) = & K_8 T_{array} V_{PV}^4 + K_9 T_{array} V_{PV}^3 + K_{10} T_{array} V_{PV}^2 + \\ & K_{11} T_{array} V_{PV} + K_{12} T_{array} \end{aligned} \qquad (2.6)$$

Desai H P et al. proposed look up table method [23], in this method, the measured values of array's voltage and current are compared with previously stored values which harmonize the operating point of array with respect to the maximum power point. The stored database contains different system condition for any insolation and temperature, and corresponding maximum power point for specific solar PV array.

The major disadvantage of this method is the requirement of bulk storage memory. Higher accuracy in tracking increases the number of operating conditions which requires more storage data. The tracking scheme is specific for array thus the implementation is complex, also considering all possible system conditions are bothersome to store and archive.

Liu X et al. discussed perturb and observe (P&O) method [24], the P&O algorithm is the most commonly used in practice because of its ease of implementation. The method is basically iterative approach, in which operating point of solar PV oscillates around the maximum power point.

The power versus voltage curve of solar PV shows that, change in power with respect to voltage (dP/dV) is positive, negative and zero for region before maximum power point, after maximum power point and at maximum power point respectively.

This method is applied by perturbing the operating voltage at regular interval and oscillating around the point dP/dV = 0 i.e., MPP. The operation explained in Table 2.1.

Table 2.1. Methodology of p&o method

Perturbation	Change in power	Next perturbation
Positive	Positive	Positive
Positive	Negative	Negative
Negative	Positive	Negative
Negative	Negative	Positive

The method is easy to implement, shows moderate accuracy, operating point oscillate around MPP, the method is slow and not suitable for fast changing condition, Oscillation can be minimized by reducing perturbation step size which slow down the MPPT, measurement of both voltage and current is required.

Femia N et al. [25], Abdelsalam A K et al. [26], Elgendy M A et al. [27], Aashoor F A O et al. [28] and Sera D et al. [29] proposed the modified methods which have the same fundamental principle of P&O with slight change in operating principle.

Hill climbing and P&O method are two different methods with same fundamental principle. P&O involves perturbation in terminal voltage to perform MPPT whereas the hill climbing method involves perturbation in duty ratio (D) Tculings W J A [30] discussed *hill climbing method* [30-32].

The methodology is explained in the Table 2.2 and flow chart given in Figure 2.4.

Table 2.2. Methodology of hill climbing method

Perturbation in terminal voltage	Change in power	Next perturbation
Positive	Positive	Positive (increment in duty ratio 'D')
Positive	Negative	Negative (decrease in duty ratio 'D')
Negative	Positive	Negative (decrease in duty ratio 'D')
Negative	Negative	Positive (increment in duty ratio 'D')

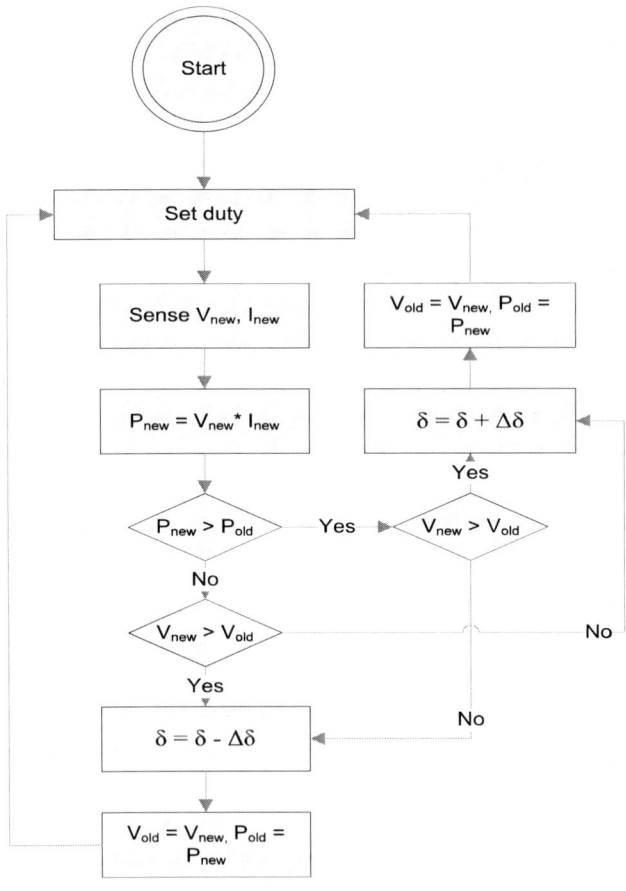

Figure 2.4. Flowchart of P&O method.

Weidong Xiao et al. proposed a modified adaptive hill climbing MPPT method for photovoltaic power systems [31] and Koutroulis E et al. [32] also proposed the modified version of the hill climbing method.

Jain S et al. [33] proposed a novel method called *beta method* in this method a coefficient beta (β) is used, which is given as;

$$\beta = \ln\frac{Ipv}{Vpv} - \left(\frac{q}{k \times T \times \eta}\right) \times Vpv \qquad (2.7)$$

Where, 'k' is Boltzmann's constant, 'η' is diode quality factor, 'T' is ambient temperature in Kelvin and 'q' is electric charge.

Eq. 2.7 indicates that value of β is independent from the insolation but depends on the temperature. In this method the solar PV operates near to this value β rather than the MPP. The method fetches the operating point close to the value of beta in few iterations thereafter P&O methods with finer steps can be used to track the exact MPP. Figure 2.5 shows the flow chart of the method.

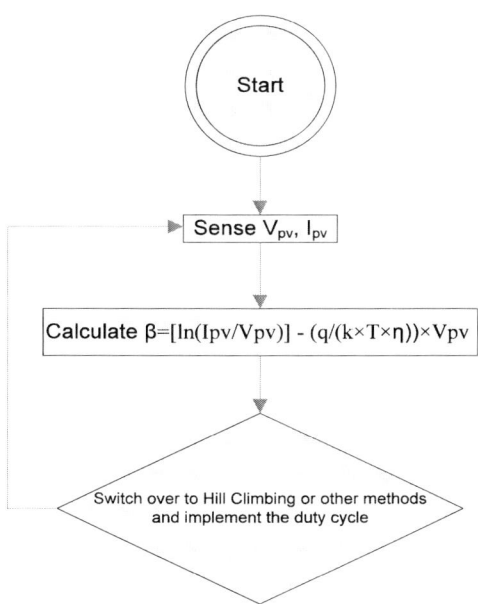

Figure 2.5. Flowchart of beta method.

Since temperature and insolation shows vague dependency, the value of β is calculated for different insolation and temperature for a particular PV system. β is evaluated with maximum and minimum value of temperature and insolation and it is observed that variation in maximum power point is small for fixed temperature even as the insolation is varied over a wide range. Despite of that there is an inverse relationship between the value of β and temperature. A range i.e., $β_{max}$ to $β_{min}$ is given for specific PV system where the algorithm gives appropriate solution. However Beta method can also be used along with another method such as incremental conductance as reported in literature.

Qiang Mei et al. proposed *Variable Step Size Incremental Resistance (INR) Method* [34]. In case of fixed step size P&O method, if step size is large method become faster but oscillations occur around MPP is higher which reduces the system efficiency and similarly smaller step size, increases the system efficiency but slows down the tracking speed. The INR method gives the solution to the problem as:

$$\delta(k) = \delta(k-1) \pm N \times \left|\frac{dP}{dV}\right| \qquad (2.8)$$

Where N is scaling factor which governs step size in case of P&O method Eq. 2.8 becomes

$$\delta(k) = \delta(k-1) \pm N \times \left|\frac{dP}{d\delta}\right| \qquad (2.9)$$

For obtaining scaling factor [34] introduced a simple method

$$N < \delta_{max} \Big/ \left|\frac{dP}{dV}\right|$$

Where δ_{max} is largest step size, INR method gives a simple and effective variable step size angle:

$$\left|\frac{dP}{dI}\right| = |tan\theta|\,; \; -90^0 < \theta < 90^0 \tag{2.10}$$

$$S_k = (\Delta I_{ref})_{max} \times sin\theta_k < (\Delta I_{ref})_{max} \tag{2.11}$$

Around MPP $sin\theta_k$ becomes lesser thus step size S_k becomes smaller.

Ansari F et al. discussed *Estimated Perturb-Perturb (EPP) Method* in their paper [35], this method is an advanced version of P&O method. EPP method uses two operating modes, *mode 1* for estimate process and *mode 2* for perturbation. The name "estimated-perturb-perturb" gives all information about the principle of this method. After two perturbations (*mode 2* in which determination of next PV voltage is done) there is one estimation mode in which controller stops tracking MPP by keeping PV voltage constant and measures only the power variation or voltage variation due to environmental changes for the next control period.

An additional estimated mode is there to improve the performance of MPPT significantly, with fast changing insolation.

Ying-Tung Hsiao et al. proposed a modified P&O method named *Three Point Weight Comparison Method* [36], the principle of this method is same as P&O method but in case of P&O, method compares only two operating point and corresponding power while in this method comparison is done regularly by perturbing the solar PV terminal voltage at three points: A, B and C. Where A is the current operating point, B is next operating point after perturbation at A and C is doubly perturbed opposite to point B.

This method is fast in comparison with P&O also suitable for fast changing condition.

Safari A et al. discussed the *Incremental Conductance (INC) Method* [37], this method is based on the fact that slop of the PV array power curve is zero at the MPP (P_{max}), This can be expressed as follows:

Power: $P = V \times I$

$$\frac{dP}{dV} = I + V\frac{dI}{dV}$$

At true MPPT $\frac{dP}{dV} = 0$

$I + V\frac{dI}{dV} = 0$

$$\frac{dI}{dV} = -\frac{I}{V} \qquad (2.12)$$

Where dI/dV: Incremental conductance,
I/V: Instantaneous conductance.

Table 2.3. Methodology of INC method

Before MPP	After MPP	At MPP
$\frac{dP}{dV} > 0$ or $\frac{dI}{dV} + \frac{I}{V} > 0$	$\frac{dP}{dV} < 0$ or $\frac{dI}{dV} + \frac{I}{V} < 0$	$\frac{dP}{dV} = 0$ or $\frac{dI}{dV} + \frac{I}{V} = 0$

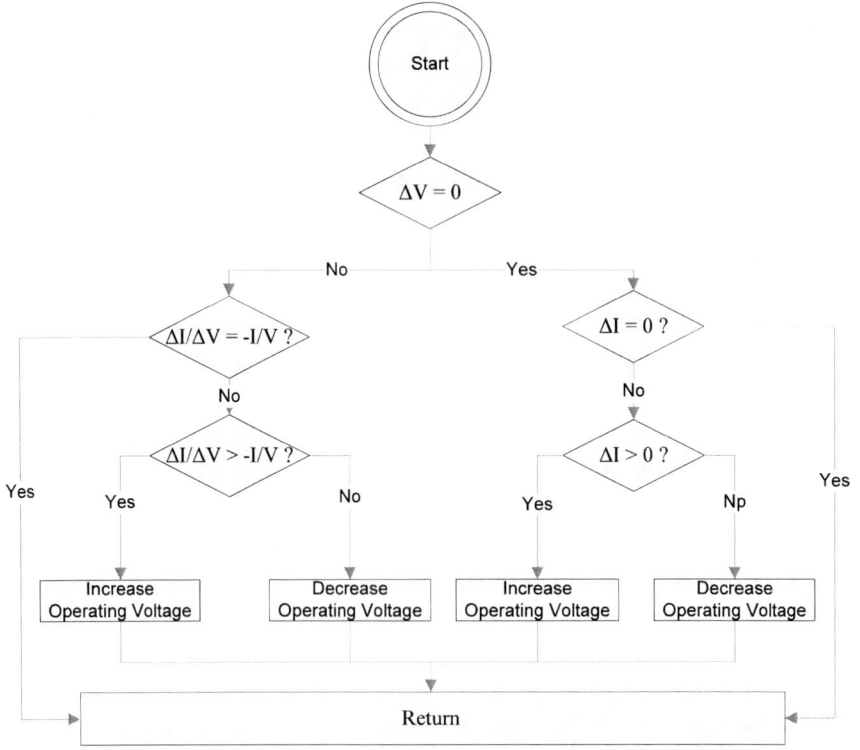

Figure 2.6. Flowchart of INC method.

Eq. 2.12 indicates that MPP can be found by comparing instantaneous conductance to the incremental conductance. The operation of this method can be divided in three zones as shown in Table 2.3. The flow chart of the method is given in Figure 2.6.

Woyte A et al. [38] also discusses the same method, the method is complex and computationally more demanding as compared to P&O.

Matsui M et al. [39] propose a novel method of MPPT *DC link capacitor droop control or Parasitic capacitance Method* based on power equilibrium at DC link, the output of the boost converter is kept constant by changing the duty ratio 'D' which is given by;

$$D = 1 - \frac{V_{cell}}{V_{link}^*} \qquad (2.13)$$

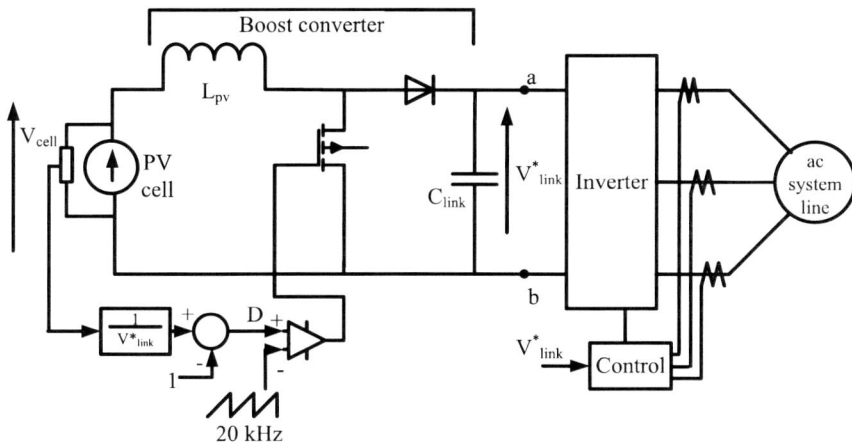

Figure 2.7. DC link capacitor droop control.

Figure 2.7 shows the schematic diagram of the method. It shows that the output of the boost converter is fed to the inverter across terminal 'a' and 'b'. Now the operating power of the array can be increased by increasing the amplitude of the line current of ac system. As long as the output power of the PV is less than the P_{max}, the power increases with the increase in line current and V^*_{link} remains constant. After reaching the maximum power the V^*_{link} starts decreasing thus the equilibrium condition is disturbed which can be controlled by the change in duty ratio of the boost converter. Thus in steady state at the constant V^*_{link} the maximum power point can be achieved.

Kitano T et al. [40] discusses the method based on same principle named *Power sensor-less MPPT control scheme* which utilize power balance at DC link-system design to ensure stability and response.

Bleijs J A M et al. [41] proposed fast maximum power point control called *dP/dV or dP/dI Feedback Control* Method. Since the power kept on increasing till the MPP and then decreases thereafter that with respect to voltage or current. In the context of the above this method compares two consecutive powers, but as compared to conventional methods, magnitude of the slope is also considered to determine MPP. Three conditions are there:

Before MPP: $p_2 > p_1$
At MPP: $p_2 = p_1$
After MPP: $p_2 < p_1$

The following formula is used to determine the error in tracking and duty ratio is adjusted accordingly.

$$\epsilon = K_c \int K_p \left(\frac{dP}{dI}\right) dt \qquad (2.14)$$

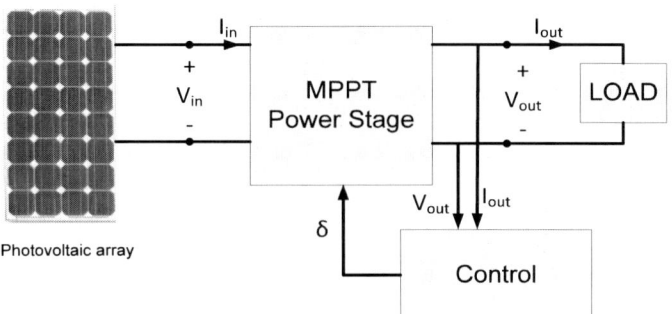

Figure 2.8. Schematic diagram of load current or load voltage maximization method.

Shmilovitz D [42] and Kislovski A S et al. [43] presents *Load Current or Load Voltage Maximization Method*, the method extracts the load parameter i.e., load voltage or load current to control the MPP instead of input current or voltage, Figure 2.8 shows the schematic diagram of the method.

MPPT power stage or a matching network has an internal controllable parameter proportional to V_{out} which controls the power flow in the network. This matching network may be loss free resistor or a transformer.

The operating principle of this method is based on the single output parameter extraction, either voltage or current. By increasing output voltage or current, power output increases until the point of maximum power, thereafter power decreases with further increase in voltage or current. In this way the operating point will converge to the MPP.

Noguchi T et al. [44], Bodur M et al. [45] and Esram Trishan [46] discussed the *Current Sweep Method*. To obtain the *I-V* characteristic of the PV array, this method uses sweep waveform for the PV array current which is updated at a regular time interval. Also for each interval V_{MPP} can then be calculated.

Yang Chen et al. [47], Femia N et al. [48] and Sreeraj E S et al. [49] discussed *One Cycle Control (OCC) Method*. One cycle control is the nonlinear control technique which is based on the integration of a variable (*voltage or current*), to convert the variable value equal to some reference value.

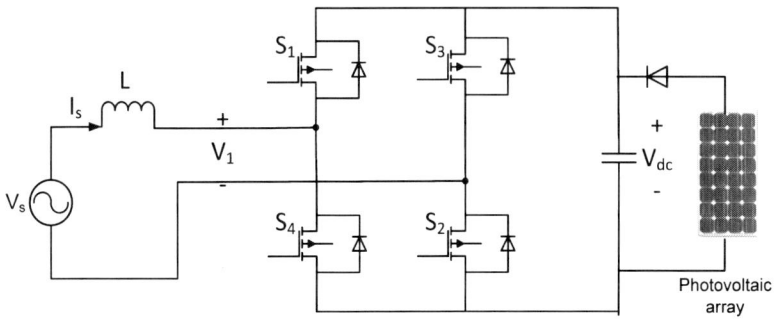

Figure 2.9. Single phase grid connected inverter with solar PV system.

OCC based method is applied to grid connected solar PV system as shown in Figure 2.9 in which single stage inverter perform the operation of MPPT.

Miao Zhang et al. [50] Il-Song Kim et al. [51] and Aghatehrani R [52] discusses *Slide mode control technique* which is used for nonlinear system, as for MPPT the application of this control technique uses two modes of operation, one is approaching mode and another one is the sliding mode, at MPP change in power with respect to voltage or current is equal to zero.

$$P = V \times I$$
$$\frac{dP}{dI} = 0$$
$$\frac{d}{dI}[I^2 R] = \left[2I \times R + I^2 \frac{d}{dI} R\right] = 0$$

This method gives the solution at MPP:

$$\sigma = 0 = 2R + I \frac{d}{dI} R \qquad (2.15)$$

Where σ is defined as sliding surface. Now duty ratio δ is updated as given in Eq. 2.16 to achieve the MPP.

$$\delta_{update} = \begin{cases} \delta + \Delta\delta \ for \ \sigma > 0 \\ \delta + \Delta\delta \ for \ \sigma < 0 \end{cases} \qquad (2.16)$$

Ho B M et al. [53] proposed *System Oscillation Method*, this method is based on the principle of maximum power transfer. It compares the ac component (oscillation due to the variation of the duty ratio) to the average value of the input voltage at the power conversion stage to determine the duty ratio. At MPP the ratio of oscillation amplitude and average voltage is constant. In this method only voltage sensor is required and methodology is easy to implement.

Bhatnagar P et al. discusses *Ripple Correlation Control (RCC)* Method [54]. This method takes the advantage of the signal ripple, which is automatically present in power converters. The ripple is interpreted as a perturbation from which an optimization can be realized. Oscillation in power provided through all pass filter which makes use of ripples to perform MPPT. Method makes use of the Eq. 2.17 that at MPP:

$$\frac{dP}{dt} \times \frac{dV}{dt} \ or \ \frac{dP}{dt} \times \frac{dI}{dt} = 0 \qquad (2.17)$$

Solodovnik E V et al. proposed *State Space based Method* [55]. The method is based on the mathematical modeling of the PV system with DC-DC converter and can be considers as a dynamic system and represented by the state space equation:

$$\dot{x}(t) = Ax(t) + B(t)u(t) + D\varepsilon(t) \qquad (2.18)$$

Where: x, state variable vector
u, switch duty ratio of DC/DC converter
ε, disturbance due to load variation
t, independent time variable

This method gives a control law 'u' to obtain MPP

$$u = K^T \times M^T g_1 + n\varepsilon(t) \qquad (2.19)$$

Where the vector K^T, M^T and the parameter n are the controller parameters that are to be defined during the design process.

The vector g_1 is the reference signal which satisfies the maximum power condition.

$$g_1 = [g(t)\dot{g}(t)\ddot{g}(t)]^T \qquad (2.20)$$

$$g(t) = V_o + i_o \left(\frac{dV_o}{dI_o}\right) \qquad (2.21)$$

V_o and i_o represent the load voltage and current.

This method ensures a globally asymptotically stable system. This method is also applicable to fast changing environmental condition.

Ching-Tsai Pan et al. proposed *Linear Current Control (LCC) or Linearization Based Method* [56]. The method graphically identifies the intersecting point of two curves, power curve of PV represented by $f(P, I) = 0$ and maximum output power curve represented by $\frac{dP}{dI} = 0$. The maximum output power curve can be represented by a linear line. To identify the intersecting point of above mentioned two curves, a simple analog circuit is used. The phenomenon is same as to find the operating point of a transistor amplifier for a specific load and the operating point is intersecting point of transistor *I-V* characteristic and load *I-V* characteristic.

Seok-Ju Lee et al. gives a novel MPPT technique named *PV Output Senseless (POS) Method* [57]. This method considers only one factor i.e., load current, to find MPP but it is applicable only for PV system with single phase DC-AC converter as shown in Figure 2.10.

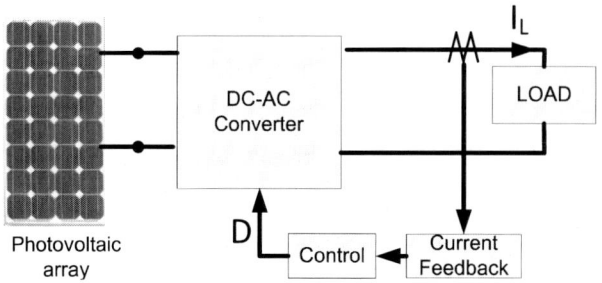

Figure 2.10. Control diagram of POS method.

As the duty ratio of DC-AC converter increases, the output current of the converter i.e., load current I_L increases but the voltage decreases thus the relation between duty ratio and voltage can be obtained.

The concept of POS control method is that, if duty ratio (D) decreases the output current of power converter or load current increases i.e., going towards the MPP. After MPP, decrease in duty ratio results decrease in current I_L i.e., going away from MPP. Thus using only one parameter load current, MPP is achieved.

Weidong Xiao et al. [58] Weidong Xiao et al. [59] and Pradhan R et al. [60] discuss *Gradient Descent or Steepest Descent Method.*

Gradient descent is the first order optimization algorithm and it is used to find a local minimum of a function by tracking steps proportional to the negative of the gradient. It is also known as steepest descent method. This method is applied to find nearest local MPP while the gradient of the function $f(V, P) = \frac{dP}{dV}$ is given by:

$$V_{k+1} = V_k + \frac{\left|\frac{dP}{dV}\right|_{V=V_k}}{k_\epsilon} \qquad (2.22)$$

Where k_ϵ is the step size corrector which decides the next step size in the direction of gradient. For MPPT operation it is required to find the solution where $\frac{dP}{dV}$ or function become minimum (zero).

Rodriguez C et al. [61] proposed a novel *Analytic Solution Based Method.* This method is based on the mean value theorem which is one of the most important method of real analysis which provides the analytic solution of a point which is closer to MPP and inside a circle of small radius ε.

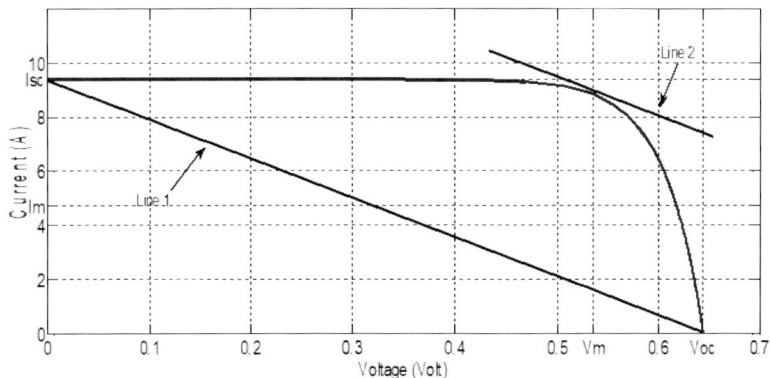

Figure 2.11. Solar cell *I-V* characteristic with parallel lines.

Line 1 in Figure 2.11 is given by:

$$i = -\frac{I_L}{V_{oc}}v + I_L \tag{2.23}$$

Where I_L is light generated current of the solar cell, and Line 2 is parallel to line 1 and tangent to the current curve.

$$\text{Now } \varepsilon > \left(v_{pv}^* + i_{pv}^* R_s\right) - \left(v_{MPP} + i_{MPP} R_s\right) \tag{2.24}$$

Where v_{pv}^*, i_{pv}^* is intersection of line 2 and near to MPP.

Longlong Zhang et al. proposed *Variable Inductance Method* [62]. This method introduces variable inductor in place of constant inductor in DC-DC buck converter along with MPPT controller. This method is robust and reliable with variation in insolation.

Use of variable inductor in place of fixed inductor in DC-DC converter reduces the overall inductor size by 75%. For high insolation current would be high, inductance of lower value is sufficient while in case of lower insolation (lower current) increased inductance required.

The minimum inductance is given by Eq. 2.25:

$$L_{min} = \frac{\delta^2(1-\delta)V_{pv}}{2f_s I_{pv}} \tag{2.25}$$

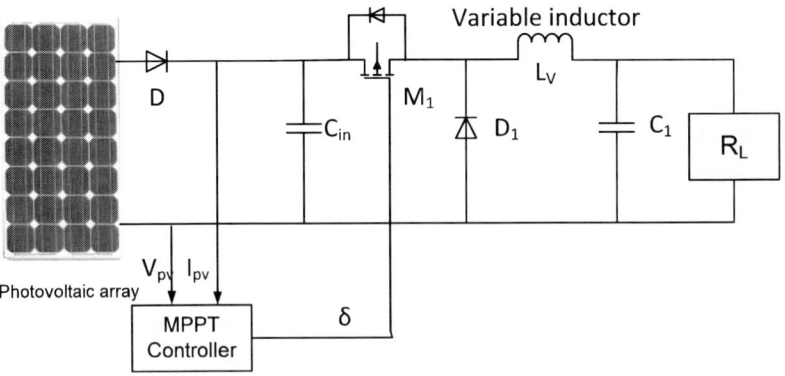

Figure 2.12. Schematic diagram of variable inductor control of MPPT.

Figure 2.12 shows the schematic diagram of variable inductor control, in which any of the conventional method can be apply with variable inductance which reduces the overall size of the inductance.

Mutoh N et al. [63], De Brito M A G et al. [64], Minwon Park et al. [65] and Coelho R F et al. [66] discussed *Temperature Based Method*. In this method temperature of solar PV is measured. Variation in MPP with respect to the temperature is obtained in similar way of constant voltage method. Eq. 26 decides the reference temperature with respect to MPP:

$$V_{MPP}(t) = V_{MPP}(T_{ref}) + T_{Kvoc}(T - T_{ref}) \qquad (2.26)$$

Where V_{MPP} is the MPP voltage, T is the working panel temperature, T_{Kvoc} is the temperature coefficient of V_{MPP}, and T_{ref} is the standard test conditions temperature.

This method is easy to implement and required simple circuitry. Voltage and temperature of PV array are required to be measured.

Peng Wang et al. [67] proposed a novel MPPT based on *Bisect Search Theorem (BST)*. It is a mathematical approach to locate the roots of any function $y = f(x)$ in an interval $[a, b]$. In context of applying BST in MPPT the function is $\frac{\Delta P}{\Delta V}$ in between the interval $[0, V_{oc}]$. As is obvious from the characteristic of solar cell, it is a function which becomes zero at MPP. Then the root $\dot{x}$ represents the solution.

Fuzzy logic soft computing technique based MPPT also found in literature. Dounis A I et al. [68], Ahmed M Othmana et al. [69], Algazar M M et al. [70], Subiyanto et al. [71], Messai A et al. [72], Bounechba H et al. [73], Patcharaprakiti et al. [74], Kharb R K et al. [75] and Altin et al. [76] discussed *Fuzzy Logic Control Method*. Fuzzy logic is a set of multiple-valued logic, as compared to binary set where variable has only two states true or false value. Fuzzy logic variable have range between zero to one, which introduces the concept of partial truth, where the variable value may range between complete true and complete false.

In the application of fuzzy logic controller for MPPT error (E) and change in error (CE) at K^{th} iteration is:

$$E(k) = \frac{P_{PV}(k) - P_{PV}(k-1)}{i_{PV}(k) - i_{PV}(k-1)} \quad (2.27)$$

$$CE(k) = E(k) - E(k-1) \quad (2.28)$$

Where P_{PV} and i_{PV} are the power and current of the PV array. In case of MPP, $E(k)$ should be zero.

For input and output variable Figure 2.13 shows the membership grades of five fuzzy subsets. The input variable like voltage and current are expressed in terms of labels (*NB*: negative big, *NS*: negative small, *ZO*: zero, *PS*: positive small, *PB*: positive big). Fuzzy controller is designed in such a way that input variable E has to be always zero, which is the condition of MPP.

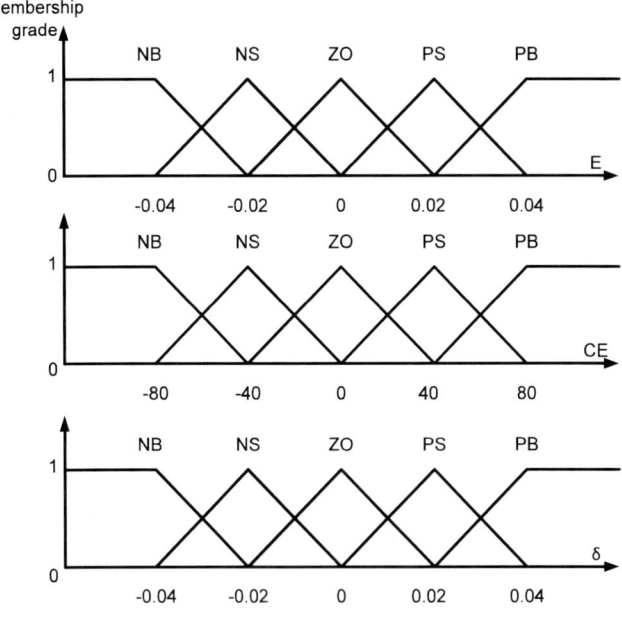

Figure 2.13. Membership function for input error change in error and duty ratio.

Artificial Neural Network soft computing technique based MPPT also found in literature. Liu Y H et al. [77], Punitha K et al. [78], Sahnoun M A et al. [79], Ocran T A et al. [80], Lin Whei-Min et al. [81], Hiyama

Takashi et al. [82], Karatepe E et al. [83] and Veerachary et al. [84] discussed Artificial Neural Network based MPPT methods. ANN is a soft computing technique inspired by central nervous system (brain) and these computational models are capable of machine learning and they are represented as the interconnected neurons (artificial nodes) to form a network similar to biological neural network.

The block diagram for application of ANN in MPPT is shown in Figure 2.14.

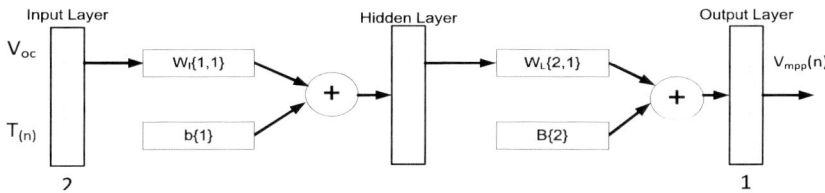

Figure 2.14. Feed forward neural network function approximator.

Two inputs are there $V_{oc}(n)$, the reference cell's open circuit voltage and time parameter $T(n)$. Training of neurons considers the connecting weights $w_I\{1,1\}$ and bias $b\{1\}$. In training process connecting weights are modified until best fit is achieved i.e., reference voltage corresponding to MPP.

Particle Swarm Optimization (PSO)/Ant colony optimization (ACO) soft computing technique based MPPT also found in literature. Ishaque K et al. [85], Khare Anula et al. [86], Miyatake M et al. [87], Lian K L et al. [88], Sundareswaran et al. [89], Ishaque Kashif et al. [90], Ishaque K et al. [91] and Liu Y H et al. [92] discussed methods based on *Particle Swarm Optimization (PSO)/Ant colony optimization (ACO)*. The PSO is a population based search algorithm, which is based on analysis of the social behavior of birds and school of fishes. The PSO approach can be applied to any optimization problem having multivariable function with multiple optimal points.

Application of PSO/ACO in MPPT is basically about the tracking in case of partial shading condition where more than one maximum point is there. Among all local maximum there is only one global maximum or

MPP. Movement of PSO agent (swarm) in search space depends on its own previous best position and the overall best position for all swarms. For every position calculation of power, is done for all agents in this way MPP achieved.

The velocity and position for next iteration for i^{th} swarm is given in Eq. 2.29 and 2.30 respectively:

$$v_i^{k+1} = wv_i^k + c_1 r_1 P_{best\ i} + c_2 r_2 g_{best} \tag{2.29}$$

$$S_i^{k+1} = s_i^k + v_i^{k+1} \tag{2.30}$$

Where v_i^{k+1} is velocity of ith swarm for iteration $k + 1$, w is learning factor, c_1, c_2 are position constant and r_1, r_2 are random numbers (their range is 0 to 1).

Some paper also found in literature which provides comparative analysis of available MPPT techniques [93-96].

Ahmed et al. [97] presents *Cuckoo search technique*, this technique is a soft computing method used to track GP by including step size of levy flight under shaded condition of PV systems. In this scheme two variables are sampled: (i) the value of operating voltage of PV array i.e., V_i (Where, i = 1,2,3 ... n) and (ii) the step size (α) selected for searching GP. The maximum PV power depends upon the fitness function(J), so J = f(V).

$$V_i^{(t+1)} = V_i^t + \alpha \oplus \text{Levy}(\lambda) \tag{2.31}$$

The levy flight generates a new voltage sample, based on the equation (41) and $\alpha = \alpha_0 (v_{best} - v_i)$. A simplified method of the levy distribution is reported in [98] as:

$$S = \alpha_0 (v_{best} - v_i) \oplus \text{Levy}(\lambda) \approx K \times \left(\frac{u}{(|v|)^{\frac{1}{\beta}}} \right) \tag{2.32}$$

Where $\beta = 1.5$, K is the multiplying coefficient of Levy (decided by manufacturer i.e., 0.05), and by using the distribution curves U and V are determined. The new maximum power that matches voltage samples are calculated. By comparing the maximum value of power, the best sample voltage is selected. According to the best sample voltage, maximum power is generated and it replaces the previous one. The process is continued until it tracks the GP.

This MPPT technique uses dc-dc buck-boost converter and is controlled by microcontroller connected to the PV system. The performance and analysis of the technique is verified on 2205W PV array experimentally and is compared by both conventional P&O and PSO techniques. The result shows that, method is efficient to track GP with fast convergence, exhibit good transient capability and higher efficiency by using two tuning parameters. The CS technique is capable of tracking the GP within 100-250 ms under different shading conditions.

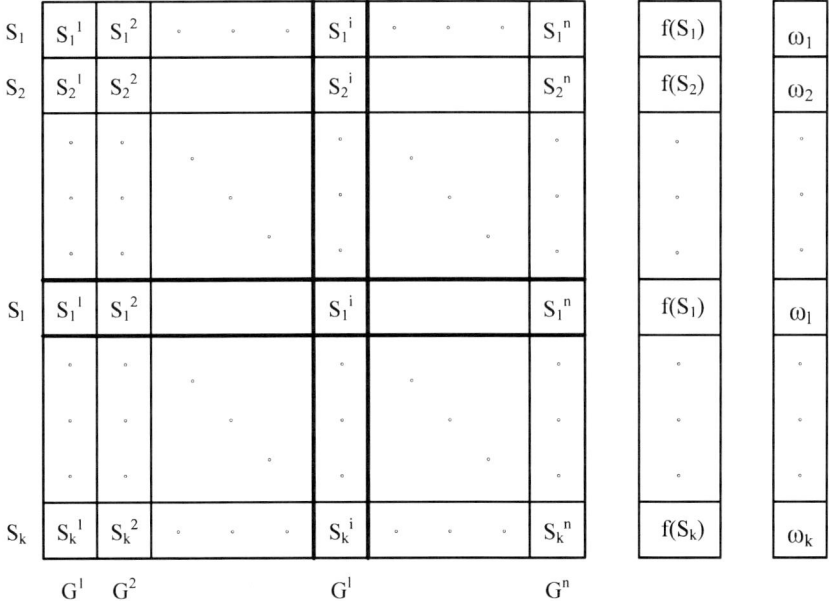

Figure 2.15. Archive solution of generation process in ACO.

Krzysztof et al. [99] present *Ant colony optimization (ACO) technique*, this technique is a probability density function algorithm in continuous domain, which is based on behavior of the ants to track GP under partial shaded condition [99-102]. Initially, K random solutions with size (K ≥ N) are generated and stored in the solution archive and parameters mean (μ), deviation (σ), and weight (w_k) for gaussian kernel is calculated, where K is random solutions as shown in Figure 2.15. The solutions are ranked according to their quality. For minimizing problem, the statement is given by:

$$f(s_1) \leq f(s_2) \leq \cdots \leq f(s_l) \leq \cdots f(s_k) \tag{2.33}$$

Each solution has a related weight w proportional to the random solution quality. Therefore:

$$(w_1) \geq (w_2) \geq \cdots \geq (w_l) \geq \cdots (w_k) \tag{2.34}$$

Gaussian function is most popular function used as a probability density function. Drawback of a single Gaussian functions is that it is unable to achieve optimal solution under multiple maximum points hence a Gaussian kernel based probability density function denoted by $G^i(x)$ for i^{th} dimension is used. It consists of several one-dimensional functions $g^i(x)$.

$$G^i(x) = \sum_{l=1}^{k} w_l g_l^i(x) = \sum_{l=1}^{k} w_l \frac{1}{\sigma_l^i \sqrt{2\pi}} \exp\left(-\frac{(x-\mu_l^i)^2}{2\sigma_l^{i^2}}\right) \tag{2.35}$$

Where μ_l, σ_l and w_l are three parameters of Gaussian kernel based probability density function i.e., mean value, standard deviation and weight of lth solution. These parameters are calculated by following equation.

$$\mu^i = \{\mu_1^i, \ldots \mu_k^i\} = \{s_1^i, \ldots s_k^i\} \tag{2.36}$$

$$\sigma_l^i = \xi \sum_{j=1}^{K} \frac{|s_j^i - s_l^i|}{K-1} \tag{2.37}$$

$$w_l = \frac{1}{QK\sqrt{2\pi}} \exp\left(-\frac{(l-1)^2}{2Q^2K^2}\right) \tag{2.38}$$

Where ξ is the convergence speed (if value of ξ is high, convergence speed is less) and Q is the parameter of best optimal solution. The small Q value gives accurately best operating solution. The probability of Gaussian function is given by

$$P_l = \frac{w_l}{\sum_{r=1}^{k} w_r} \tag{2.39}$$

The iteration process is repeated until new optimal solution (M) is produced. To archive best optimal solution, the total solutions (M+K) are ranked again. The performance and analysis of this technique is compared with P&O and constant voltage tracking MPPT techniques under simulation and the result shows higher efficiency. This technique is efficient to track the GP under various shading condition.

2.3. Comparison of MPPT Methods

Comparison of MPPT methods is done on the basis of following 13 factors in this chapter:

1. Category
2. Dependency of PV array
3. Implementation methodology
4. Sensor required
5. Stages of energy conversion
6. Partial shading enabled
7. Grid interaction
8. Analog or Digital
9. Tracking efficiency
10. Tracking speed
11. Cost

12. Size of the PV array
13. Product available in market.

2.3.1. Category

MPPT methods can be classified under Indirect Control (IND), Direct Control (DIR) and soft computing techniques (SC) on the basis of their control strategy.

2.3.1.1. Indirect Control Methods

Indirect control methods are typically based on mathematical relationship obtained from the empirical data which may include the parameters and characteristic of the solar PV panel. Maximum power points are basically predicted offline using various algorithms, empirical data or mathematical equations. These methods are not suitable with fast changing environmental condition and partial shading condition, also the MPP given by the method is not true but it is an approximation based on the database. Example: Constant voltage, constant current, look up table, curve fitting, pilot cell methods.

2.3.1.2. Direct Control Methods

Direct control methods are search algorithms which locate the maximum power points against changing atmospheric conditions online. They are generally based on the sampling based control or modulation based control strategy. Example: Hill Climbing, P&O and INC etc.

2.3.1.3. Soft Computing Technique Based Methods

Methods based on the soft computing techniques belong to this category. Example: methods based on the Genetic algorithm, Artificial Neural Network and particle swarm optimization.

2.3.2. Dependency of PV Array (DPVA)

This category explains the state of dependency (whether dependent or independent) of MPPT methods on the type and size of PV system. Example: Constant voltage method is PV system specific.

2.3.3. Implementation Methodology

This category explains the degree of complexity of the circuitry required for implementation of the method. Example: constant voltage method is less complex as compared to the P&O method.

2.3.4. Sensor Required

The control parameter or sensed parameter required to find out the point of maximum power such as Voltage (V), Current (I), Temperature (T) or may be combination of these three.

2.3.5. Stages of Energy Conversion

In solar PV systems energy conversion stages such as DC-DC, DC-AC or both, required to control the output of source with respect to maximum power.

2.3.6. Partial Shading Enabled (PSE)

Under partial shading condition the output of the PV shows multiple local maxima. This category defines the compatibility of the method to find global maximum [103-110].

2.3.7. Grid Interaction (GI)

Methods can also be classified on the basis of their mode of connection with the grid such as grid connected or off grid.

2.3.8. Analog or Digital

Methods can be classified on the basis of the kind of operating circuitry required such as analog (A), digital (D), combination of both.

2.3.9. Tracking Efficiency (TE)

Tracking efficiency basically defines the tracking accuracy of the tracker. A MPPT tracker can be efficient upto100% if it delivers maximum power which is available. The efficiency of tracker is given as;

$$T.E. = \eta = \frac{P_{Out}}{P_{MPP}} \times 100$$

In this paper tracking efficiency is evaluated in terms of three different categories such as good, medium and poor.

Example:

- Methods based on the soft computing techniques are more accurate hence their tracking efficiency is good.
- P&O method gives medium efficiency.
- Constant current method, constant voltage method and pilot cell method give poor efficiency (as compared to conventional P&O method).

2.3.10. Tracking Speed

Tracking speed of the MPPT tracker defines the speed of the tracker to achieve the MPP. Example: in case of P&O method increase in perturbation size results increase in Tracking Speed but the accuracy or Tracking efficiency becomes poor.

2.3.11. Cost

This category defines economy of the MPPT tracker in comparison with the other trackers. The cost of any tracking scheme is directly related with sensors required, computational procedure and the circuitry used to implement the same. The methodology used to evaluate the cost of tracking scheme is based on above mentioned parameters. Methods requiring complex circuitry are more costly. In this paper cost of tracking scheme is separated in four categories that is inexpensive (INEX), medium (Med.), expensive (EX) and very expensive (V.EX).

Example:

- Constant voltage method and constant current method require less computation to reach MPP. Only one sensor is required with less complex implementation circuitry, hence can be categorized in inexpensive category.
- P&O and INC require two sensors and more computation to obtain MPP as compared to constant current and constant voltage so these methods are categorized in medium category.
- Soft computing techniques require more complex hardware thus are very expensive.

2.3.12. Size of the PV Array

Size of the PV array is another important factor and can be split in three categories as small systems(S) which are less than 1 KW, medium systems (M) which are in the range of 1 KW- 1 MW and large systems (L) which are above 1 MW.

Example:

- Less expensive MPPT techniques are preferred for small size systems as expensive MPPT will result in increased cost of overall system with a little increment in power.
- For medium system P&O and INC is suitable.
- For large systems accurate method is preferred as the power increments due to MPPT tracker is significant and partial shading effect is a major concern.

2.3.13. Product Available in Market

On the basis of the operating principle of the method, available commercial products in the market are comprised in this category. A survey on available commercial products is done by means of product datasheet analysis and conversation with the manufacturer for evaluation of this category [108-112].

Table 2.4. Comparison of the MPPT methods

SN	Method	Category	DP VA	Implementation of methodology	Sensor required V	Sensor required I	Sensor required T	Stages of energy conversion	PSE	GI	Analog or Digital	T.E.	Tracking speed	Cost	Size	Product available in market
1	Constant voltage	IND		Simple				DC-DC			A,D	Poor	Fast	INEX	S	
2	Pilot cell	IND		Simple				DC-DC			A,D	Poor	Fast	INEX	S	
3	Constant Current	IND		Simple				DC-DC			A,D	Poor	Fast	INEX	S	
4	Curve Fitting	IND		Simple				DC-DC			D	Med.	Fast	INEX	S	OutBack Power, USA-FLEXmax 80,FLEXmax60
5	Look Up Table	IND		Simple				DC-DC			D	Med.	Fast	INEX	S	
6	Perturb and Observe	DIR		Simple				Both			A,D	Good	Slow	Med.	S/M	Genasun, USA GV Boost charge controller with MPPT; Blue Chip Energy Solution Pvt. Ltd., India- Solar MPPT Charge controller
7	Hill Climbing	DIR		Simple				Both			A,D	Good	Slow	Med.	S/M	
8	Beta	DIR		Complex				Both			D	Good	Fast	EX	M/L	
9	Variable Step Size Incremental Resistance	DIR		Complex				Both			D	Good	Fast	EX	M	

S N	Method	Category	DP VA	Implementation of methodology	Sensor required V	Sensor required I	Sensor required T	Stages of energy conversion	PSE	GI	Analog or Digital	T.E.	Tracking speed	Cost	Size	Product available in market
10	Estimated Perturb-Perturb	DIR		Complex				Both			A,D	Good	Med.	EX	M	
11	Three Point Weight Comparison	DIR		Complex				Both			A,D	Good	Slow	EX	M	
12	Incremental Conductance	DIR		Med.				Both			D	Good	Med.	Med.	S/M	
13	Parasitic capacitance	DIR		Complex				Both			D	Good	Med.	EX	M	
14	dP/dV or dP/dI Feedback Control	DIR		Complex				DC-DC			A,D	Good	Med.	EX	M	
15	Load Current or Load Voltage Maximization	DIR		Med.				DC-AC			A	Good	Slow	Med.	S/M	
16	Current Sweep	DIR		Complex				DC-DC			D	Med.	Med.	Med.	S/M	Blue Sky Energy, USA New-Solar Boost 300i, 2512i-HV & 2512ix-HV
																Morningstar Corporation, USA TriStar MPPT & SunSaver MPPT

Table 2.4. (Continued)

SN	Method	Category	DP	VA	Implementation of methodology	Sensor required V	Sensor required I	Sensor required T	Stages of energy conversion	PSE	GI	Analog or Digital	T.E.	Tracking speed	Cost	Size	Product available in market
17	One Cycle Control	DIR			Med.				DC-AC			A,D	Good	Fast	Med.	S/M	
18	Slide Control	DIR			Med.				Both			D	Good	Fast	EX	M	
19	System Oscillation	DIR			Complex				DC-DC			A	Good	Med.	EX	M	
20	Ripple Correlation Control	DIR			Complex				DC-DC			A	Good	Fast	EX	M	
21	State Space based	DIR			Complex				DC-DC			D	Good	Med.	EX	M	
22	Linear Current Control	DIR			Simple				DC-DC			D	Good	Fast	EX	M	
23	PV Output Senseless	DIR			Simple				DC-DC			D	Good	Med.	EX	M	
24	Gradient Descent	DIR			Complex				DC-DC			D	Good	Med.	EX	M	
25	Analytic Solution Based	DIR			Complex				DC-DC			A,D	Good	Med.	EX	M	
26	Variable Inductance	DIR			Med.				Both			A,D	Med.	Med.	EX	M	
27	Temperature Based	SC			Med.				DC-DC			A	Good	Med.	EX	M	

S N	Method	Category	DP	VA	Implementation of methodology	Sensor required			Stages of energy conversion	PSE	GI	Analog or Digital	T.E.	Tracking speed	Cost	Size	Product available in market
						V	I	T									
28	Bisect Search Theorem	SC			Complex				Both			D	Good	Med.	V. EX	L	
29	Fuzzy Logic Control	SC			Complex				Both			D	V. Good	Fast	V. EX	L	
30	Artificial Neural Network (ANN) based	SC			Complex				Both			D	V. Good	Fast	V. EX	L	
31	Particle Swarm Optimization	SC			Complex				Both			D	V. Good	Fast	V. EX	L	
32	Cuckoo search technique	SC			Complex				Both			D	V. Good	Fast	V. EX	L	
33	Ant colony optimization (ACO) technique	SC			Complex				Both			D	V. Good	Fast	V. EX	L	

Most of the manufacturer uses their own MPP methods for example Steca Elektronic, Germany uses unique MPP tracking for products Steca Solarix MPPT 1010 and 2010 [113].

Table 2.4 gives the comparative analysis of the MPPT methods.

2.4. Comparative Analysis of the MPPT Methods

Table 2.4 gives the brief comparison of MPPT methods and shows that all methods have their own advantages and disadvantages. Some of the methods show very effective results such as soft computing technique but the methodologies used were complicated. The methods which are simple in implementation such as constant voltage method, constant current method and pilot cell method are less accurate. Perturb and observe method is commonly used method because its implementation circuitry is not complex but it shows sluggish response where environmental conditions changes rapidly. Beta method shows proficient results with fast change in insolation only if variation in temperature is less. Some methods are based on mathematical optimization algorithms to solve the non linear problem such as state space based method, analytic solution based method and steepest descent or gradient descent method, they required more computation thus the complexity increases and in case of partial shading tracking efficiency goes down. In partial shading condition soft computing techniques such as ANN based method, GA based method and PSO based method give conspicuous tracking efficiency but overall system becomes expensive.

CONCLUSION

This chapter presents a literature survey of 33 MPPT techniques and their classification on the basis of 13 parameters. All important methods are broadly classified and found to have variations in either implied algorithm or in hardware architecture. It is really difficult to select application specific MPPT technique as all reported methods have its own merits and demerits. The important methods in a group are therefore compared against vital parameters and study is lucidly presented in tabular form for ready apprehension of users. Few important parameters considered for comparison include number of sensors used, speed, complexity etc. The recent development based on current and voltage equalization scheme are also reviewed and compared, which are seldom found in other reviews. This survey should serve as valuable reference for all researchers' in the field.

REFERENCES

[1] Chauhan Anurag, R P Saini, "A review on Integrated Renewable Energy System based power generation for stand-alone applications, Configurations, storage options, sizing methodologies and control," *Renewable and Sustainable Energy Reviews* 2014, vol. 38, pp. 120-99.

[2] Sebri Maamar, Ousama Ben Salha, "On the causal dynamics between economic growth, renewable energy consumption, CO_2 emissions and trade openness, Fresh evidence from BRICS countries," *Renewable and Sustainable Energy Reviews* 2014, vol. 39, pp. 23-14.

[3] International Energy Agency, *Tracking Clean Energy Progress*, Paris, France 2014.

[4] International Energy Agency, *World Energy Investment Outlook - Special Report Paris,* France 2014.

[5] International Energy Agency, "*World energy outlook. Paris,*" France 2013.

[6] Mads Troldborg, Simon Heslop, Rupert L Hough, "Assessing the sustainability of renewable energy technologies using multi-criteria analysis, Suitability of approach for national-scale assessments and associated uncertainties," *Renewable and Sustainable Energy Reviews* 2014, vol. 39, pp. 1184-1173.

[7] Benson Christopher L, Christopher L Magee. "On improvement rates for renewable energy technologies, Solar PV, wind turbines, capacitors, and batteries," *Renewable Energy* 2014, vol. 68, pp. 751-745.

[8] Lesourd Jean-Baptiste. "Solar photovoltaic systems, the economics of a renewable energy resource," *Environmental Modelling & Software* 2001, vol. 16 no.2, pp. 156-147.

[9] Dincer Furkan. "The analysis on photovoltaic electricity generation status, potential and policies of the leading countries in solar energy," *Renewable and Sustainable Energy Reviews* 2011, vol. 15.1, pp. 720-713.

[10] Parida Bhubaneswari, S Iniyan, Ranko Goic. "A review of solar photovoltaic technologies," *Renewable and sustainable energy reviews* 2011, vol. 15.3, pp. 1625-1636.

[11] Singh Solanki Chetan. *Solar photovoltaics, fundamentals, technologies and applications*, PHI Learning Pvt. Ltd., 2011.

[12] Yi Huang, Fang Z. Peng, "Survey of the power conditioning system for pv power generation," *Power Electronics Specialists Conference,* 2006.

[13] *PhD Thesis of P. Bhatnagar,* MANIT, Bhopal, India, Dec-2014.

[14] *PhD Thesis of S. Nema,* RGTU Bhopal, India, 2012.

[15] *PhD Thesis of R. K. Nema,* Barkatullah University, 2004.

[16] Kavadias K.A. "2.19 - Stand-Alone, Hybrid Systems," In *Comprehensive Renewable Energy,* edited by Ali Sayigh, Elsevier, Oxford, 2012, pp. 623-655.

[17] Yanine Franco F, Enzo E. Sauma, "Review of grid-tie micro-generation systems without energy storage, Towards a new approach to sustainable hybrid energy systems linked to energy efficiency," *Renewable and Sustainable Energy Reviews,* October 2013, vol. 26, pp. 60-95.

[18] Deepak Paramashivan Kaundinya, P. Balachandra, N.H. Ravindranath, "Grid-connected versus stand-alone energy systems for decentralized power—A review of literature," *Renewable and Sustainable Energy Reviews,* vol. 13, no. 8, pp. 2041-2050.

[19] Leedy A W, Liping Guo, Aganah K A, "A constant voltage MPPT method for a solar powered boost converter with DC motor load," *Proceedings of IEEE Southeastcon* 2012, vol. 1.6, pp. 18-15.

[20] Salameh Z M, Dagher F, & Lynch W A, "Step-down maximum power point tracker for photovoltaic systems," *Solar Energy* 1991, vol. 46(5), pp. 279-282.

[21] Alghuwainem S M, "Matching of a dc motor to a photovoltaic generator using a step-up converter with a current-locked loop," *IEEE Transactions on Energy Conversion*, 1994, vol. 9(1), pp. 192-198.

[22] Leedy A W, Garcia K E, "Approximation of P-V characteristic curves for use in maximum power point tracking algorithms," *System Theory (SSST) 45th Southeastern Symposium* 2013, pp. 93-88.

[23] Desai H P, Patel H K, "Maximum Power Point Algorithm in PV Generation, An Overview," In *Power Electronics and Drive Systems PEDS 7th International Conference* 2007, pp. 630-624.

[24] Liu X, Lopes L A C. "An improved perturbation and observation maximum power point tracking algorithm for PV arrays," *Power Electronics Specialists IEEE 35th Annual Conference*, 3, 2010-2005.

[25] Femia N, Petrone G, Spagnuolo G, Vitelli M, "Optimization of perturb and observe maximum power point tracking method," *IEEE Transactions on Power Electronics* 2005, vol. 20, pp. 973-963.

[26] Abdelsalam A K, Massoud A M, Ahmed S, Enjeti P, "High-Performance Adaptive Perturb and Observe MPPT Technique for Photovoltaic-Based Microgrids," *IEEE Transactions on Power Electronics* 2011, vol. 26,4, pp. 2021-1010.

[27] Elgendy M A, Zahawi B, Atkinson D J, "Assessment of Perturb and Observe MPPT Algorithm Implementation Techniques for PV Pumping Applications," *IEEE Transactions on Sustainable Energy* 2012, vol. 3,1, pp. 33-21.

[28] Aashoor F A O, Robinson FVP, "A variable step size perturb and observe algorithm for photovoltaic maximum power point tracking," Universities *Power Engineering 47th International Conference (UPEC)* 2012,1,6, pp. 7-4.

[29] Sera D, Mathe L, Kerekes T, Spataru S V, Teodorescu R, "Perturb-and-Observe and Incremental Conductance MPPT Methods for PV Systems," *IEEE Journal of Photovoltaics* 2013, vol. 3,3, pp. 1078-1070.

[30] Teulings W J A, Marpinard J C, Capel A, O'Sullivan D, "A new maximum power point tracking system," *Power Electronics Specialists 24th Annual IEEE Conference PESC* 1993, pp. 838-833.

[31] Weidong Xiao, Dunford W G, "A modified adaptive hill climbing MPPT method for photovoltaic power systems," *Power Electronics Specialists IEEE 35th Annual Conference* 2004, vol. 3, pp. 1963-1957.

[32] Koutroulis E, Kalaitzakis K, Voulgaris N C, "Development of a microcontroller-based, photovoltaic maximum power point tracking control system," *IEEE Transactions on Power Electronics* 2001, vol. 16,1, pp. 54-46.

[33] Jain S, Agarwal V, "A new algorithm for rapid tracking of approximate maximum power point in photovoltaic systems," *IEEE Power Electronics Letters* 2004, vol. 2,1, pp. 19-16.

[34] Qiang Mei, Mingwei Shan, Liying Liu, Guerrero J M, "A Novel Improved Variable Step-Size Incremental-Resistance MPPT Method for PV Systems," *IEEE Transactions on Industrial Electronics* 2011, vol. 58,6, pp. 2434-2427.

[35] Ansari F, Iqbal A, Chatterji S, Afzal A, "Control of MPPT for photovoltaic systems using advanced algorithm EPP," *International Conference on Power Systems ICPS* 2009,1,6, pp. 29-27.

[36] Ying-Tung Hsiao, China-Hong Chen, "Maximum power tracking for photovoltaic power system," *37th IAS Annual Meeting Industry Applications Conference* 2002, vol. 2, pp. 1040-1035.

[37] Safari A, Mekhilef S, "Simulation and Hardware Implementation of Incremental Conductance MPPT With Direct Control Method Using Cuk Converter," *IEEE Transactions on Industrial Electronics* 2011, vol. 58,4, pp. 1161-1154.

[38] Woyte A, Van Thong Vu, Belmans R, Nijs J, "Voltage fluctuations on distribution level introduced by photovoltaic systems," *IEEE Transactions on Energy Conversion* 2006, vol. 21,1, pp. 209-202.

[39] Matsui M, Kitano T, De-hong Xu, Zhong-qing Yang, "A new maximum photovoltaic power tracking control scheme based on power equilibrium at DC link," *IEEE Industry Applications Conference, Conference Record of the Thirty-Fourth IAS Annual Meeting* 1999,2, pp. 809-804.

[40] Kitano T, Matsui M, De-hong Xu, "Power sensor-less MPPT control scheme utilizing power balance at DC link-system design to ensure stability and response," *The 27th Annual Conference of the IEEE Industrial Electronics Society IECON* -01 2001, vol. 2, pp. 1314-1309.

[41] Bleijs J A M, Gow J A, "Fast maximum power point control of current-fed DC-DC converter for photovoltaic arrays," *Electronics Letters* 2001, vol. 1, pp. 6-5.

[42] Shmilovitz D, "On the control of photovoltaic maximum power point tracker via output parameters," *IEE Proceedings Electric Power Applications* 2005, vol. 152,2, pp. 248-239.

[43] Kislovski A S, Redl R, "Maximum-power-tracking using positive feedback," *Power Electronics Specialists 25th Annual IEEE Conference PESC*-1994, vol. 2, pp. 1068-1065.

[44] Noguchi T, Matsumoto H, "Maximum-power-point tracking method of photovoltaic power system using single transducer," *The 29th Annual Conference of the IEEE Industrial Electronics Society, IECON* -2003, vol. 3, pp. 2355-2350.

[45] Bodur M, Ermis M, "Maximum power point tracking for low power photovoltaic solar panels," *Electrotechnical 7th Mediterranean Conference* 1994,2, pp. 761-758.

[46] Esram Trishan, Patrick L Chapman. "Comparison of photovoltaic array maximum power point tracking techniques" *IEEE transactions on energy conversion EC* 2007, vol. 22.2, pp. 449-439.

[47] Yang Chen, Smedley K M, "A cost-effective single-stage inverter with maximum power point tracking," *IEEE Transactions on Power Electronics* 2004, vol. 19,5, pp. 1294-1289.

[48] Femia N, Granozio D, Petrone G, Spagnuolo G, Vitelli M, "Optimized one-cycle control in photovoltaic grid connected applications," *IEEE Transactions on Aerospace and Electronic Systems* 2006, vol. 42,3, pp. 972-954.

[49] Sreeraj E S, Chatterjee K, Bandyopadhyay S, "One-Cycle-Controlled Single-Stage Single-Phase Voltage-Sensorless Grid-Connected PV System," *IEEE Transactions on Industrial Electronics* 2013, vol. 60,3, pp. 1224-1216.

[50] Miao Zhang, Jie Wu, Hui Zhao, "The application of slide technology in PV maximum power point tracking system," *Fifth World Congress on Intelligent Control and Automation WCICA* 2004, vol. 6, pp. 5594-5591.

[51] Il-Song Kim, Myung-Bok Kim, Myung-Joong Youn, "New Maximum Power Point Tracker Using Sliding-Mode Observer for Estimation of Solar Array Current in the Grid-Connected Photovoltaic System," *IEEE Transactions on Industrial Electronics* 2006, vol. 53,4, pp. 1035-1027.

[52] Aghatehrani R, Kavasseri R, "Sensitivity-Analysis-Based Sliding Mode Control for Voltage Regulation in Microgrids," *IEEE Transactions on Sustainable Energy* 2013, vol. 4,1, pp. 57-50.

[53] Ho B M, Chung, H S, & Lo, W L, "Use of system oscillation to locate the MPP of PV panels," *Power Electronics Letters IEEE*, 2004, vol. 2(1), pp. 1-5.

[54] Bhatnagar Pallavee, R K Nema, "Maximum power point tracking control techniques, State-of-the-art in photovoltaic applications," *Renewable and Sustainable Energy Reviews* 2013, vol. 23, pp. 241-224.

[55] Solodovnik E V, Liu Shengyi Dougal R A, "Power controller design for maximum power tracking in solar installations," *IEEE Transactions on Power Electronics* 2004, vol. 19,5, pp. 1304-1295.

[56] Ching-Tsai Pan, Jeng-Yue Chen, Chin-Peng Chu, Yi-Shuo Huang, "A fast maximum power point tracker for photovoltaic power systems," *The 25th Annual Conference of the IEEE Industrial Electronics Society IECON* 1999,1, pp. 393-390.

[57] Seok-Ju Lee, Hae-Yong Park, Gyeong-Hun Kim, Hyo-Ryong Seo, Ali M H, Minwon Park, et al., "The experimental analysis of the grid- connected PV system applied by POS MPPT," *International Conference on Electrical Machines and Systems ICEMS* 2007, pp. 1791-1786.

[58] Weidong Xiao, Lind M G J, Dunford W G, Capel A, "Real-Time Identification of Optimal Operating Points in Photovoltaic Power Systems," *IEEE Transactions on Industrial Electronics* 2006,53,4, pp. 1026-1017.

[59] Weidong Xiao, Dunford W G, Palmer P R, Capel A, "Application of Centered Differentiation and Steepest Descent to Maximum Power Point Tracking," *IEEE Transactions on Industrial Electronics* 2007, vol. 54,5, pp. 2549-2539.

[60] Pradhan R, Subudhi B, "A steepest-descent based maximum power point tracking technique for a photovoltaic power system. Power," *2nd International Conference on Control and Embedded Systems (ICPCES)* 2012, vol. 1,6, pp. 19-17.

[61] Rodriguez C, Amaratunga G, "Analytic Solution to the Photovoltaic Maximum Power Point Problem," *IEEE Transactions on Circuits and Systems I Regular Papers* 2007,54,9, pp. 2060-2054.

[62] Longlong Zhang, Hurley W G, Wölfle W H, "A New Approach to Achieve Maximum Power Point Tracking for PV System With a Variable Inductor," *IEEE Transactions on Power Electronics* 2011, vol. 26,4, pp. 1037-1031.

[63] Mutoh N, Matuo T, Okada K, Sakai M, "Prediction-data-based maximum-power-point-tracking method for photovoltaic power generation systems," *Power Electronics Specialists IEEE 33rd Annual Conference* 2002,3, pp. 1494-1489.

[64] De Brito M A G, Galotto L, Sampaio L P, de Azevedo e Melo G, Canesin C A, "Evaluation of the Main MPPT Techniques for

Photovoltaic Applications," *IEEE Transactions on Industrial Electronics* 2013, vol. 60,3, pp. 1167-1156.

[65] Minwon Park, In-Keun Yu, "A study on the optimal voltage for MPPT obtained by surface temperature of cell," *30th Annual Conference of IEEE Industrial Electronics Society IECON* 2004,3, pp. 2045-2040.

[66] Coelho R F, Concer F M, Martins D C, "A MPPT approach based on temperature measurements applied in PV systems" *9th IEEE/IAS International Conference on Industry Applications (INDUSCON)* 2010, pp. 6-1.

[67] Peng Wang, Haipeng Zhu, Weixiang Shen, Fook Hoong Choo, Poh Chiang Loh, Kuan Khoon Tan, "A novel approach of maximizing energy harvesting in photovoltaic systems based on bisection search theorem," *Applied Power Electronics Twenty-Fifth Annual IEEE Conference and Exposition* (APEC) 2010, pp. 2148-2143.

[68] Dounis A I, Kofinas P, Alafodimos C, Tseles D, "Adaptive fuzzy gain scheduling PID controller for maximum power point tracking of photovoltaic system," *Renewable Energy* 2013, vol. 60, pp. 214-202.

[69] Othmana Ahmed M, Mahdi M M El-arinia, Ahmed Ghitasb, Ahmed Fathya, "Realworld maximum power point tracking simulation of PV system based on Fuzzy Logic control," *NRIAG Journal of Astronomy and Geophysics* 2012, vol. 1.2, pp. 194-186.

[70] Algazar M M, Al-Monier H, EL-Halim H A, Salem M E E K, "Maximum power point tracking using fuzzy logic control," *International Journal of Electrical Power & Energy Systems* 2012, vol. 39,1, pp. 28-221.

[71] Subiyanto Subiyanto, Azah Mohamed, Hannan M, "Intelligent maximum power point tracking for PV system using Hopfield neural network optimized fuzzy logic controller," *Energy and Buildings* 2012, vol. 51, pp. 38-29.

[72] Messai A, Mellit A, Guessoum A, Kalogirou S A, "Maximum power point tracking using a GA optimized fuzzy logic controller and its FPGA implementation," *Solar energy* 2011, vol. 85,2, pp. 277-265.

[73] Bounechba H, Bouzid A, Nabti K, Benalla H, "Comparison of Perturb & Observe and Fuzzy Logic in Maximum Power Point Tracker for PV Systems," *Energy Procedia* 2014, vol. 50, pp. 684-677.

[74] Patcharaprakiti Nopporn, Suttichai Premrudeepreechacharn, Yosanai Sriuthaisiriwong, "Maximum power point tracking using adaptive fuzzy logic control for grid-connected photovoltaic system," *Renewable Energy* 2005, vol. 30,11, pp. 1788-1771.

[75] Kharb R K, Shimi S L, Chatterji S, Ansari M F, "Modeling of solar PV module and maximum power point tracking using ANFIS," *Renewable and Sustainable Energy Reviews* 2014, vol. 33, pp. 612-602.

[76] Altin, Necmi, Saban Ozdemir, "Three-phase three-level grid interactive inverter with fuzzy logic based maximum power point tracking controller," *Energy Conversion and Management* 2013, vol. 69, pp. 26-17.

[77] Liu Y H, Liu C L, Huang J W, Chen J H, "Neural-network-based maximum power point tracking methods for photovoltaic systems operating under fast changing environments," *Solar Energy* 2013, vol. 89, pp. 53-53.

[78] Punitha K, Devaraj D, Sakthivel S, "Artificial neural network based modified incremental conductance algorithm for maximum power point tracking in photovoltaic system under partial shading conditions," *Energy* 2013,62, pp. 340-330.

[79] Sahnoun M A, Ugalde H M R, Carmona J C, Gomand J, "Maximum Power point Tracking Using P&O Control Optimized by a Neural Network Approach, A Good Compromise between Accuracy and Complexity," *Energy Procedia* 2013, vol. 42, pp. 659-650.

[80] Ocran T A, Cao J, Cao B, Sun X, "Artificial neural network maximum power point tracker for solar electric vehicle," *Tsinghua Science & Technology* 2005, vol. 10,2, pp. 208-204.

[81] Lin Whei-Min, Chih-Ming Hong, Chiung-Hsing Chen, "Neural-network-based MPPT control of a stand-alone hybrid power

generation system," *IEEE Transactions on Power Electronics* 2011, vol. 26,12, pp. 3581-3571.

[82] Hiyama Takashi, Shinichi Kouzuma, Tomofumi Imakubo, "Identification of optimal operating point of PV modules using neural network for real time maximum power tracking control," *IEEE transactions on Energy conversion* 1995, vol. 10,2, pp. 367-360.

[83] Karatepe E, T Hiyama, "Artificial neural network-polar coordinated fuzzy controller based maximum power point tracking control under partially shaded conditions," *Renewable Power Generation IET* 2009, vol. 3,2, pp. 253-239.

[84] Veerachary Mummadi, Tomonobu Senjyu, Katsumi Uezato, "Neural-network-based maximum-power-point tracking of coupled-inductor interleaved-boost-converter-supplied PV system using fuzzy controller," *IEEE Transactions on Industrial Electronics* 2003, vol. 50,4, pp. 758-749.

[85] Ishaque K, Salam Z, Shamsudin A, Amjad M, "A direct control based maximum power point tracking method for photovoltaic system under partial shading conditions using particle swarm optimization algorithm," *Applied Energy* 2012, vol. 99, pp. 422-414.

[86] Khare Anula, Saroj Rangnekar, "A review of particle swarm optimization and its applications in Solar Photovoltaic system," *Applied Soft Computing* 2013, vol. 13,5, pp. 3006-2997.

[87] Miyatake M, Toriumi F, Fujii N, Ko H, "Maximum power point tracking of multiple photovoltaic arrays, a PSO approach," *IEEE Transactions on Aerospace and Electronic Systems* 2011, vol. 47,1, pp. 380-367.

[88] Lian K L, J H Jhang, I S Tian, "A Maximum Power Point Tracking Method Based on Perturb-and-Observe Combined With Particle Swarm Optimization," *IEEE Journal of Photovoltaics* 2014, vol. 4,2, pp. 8-1.

[89] Sundareswaran Kinattingal, Sankar Peddapati, S Palani, "Application of random search method for maximum power point

tracking in partially shaded photovoltaic systems," *IET Renewable Power Generation* 2014, vol. 8,6, pp. 678-670.

[90] Ishaque Kashif, Zainal Salam, "A deterministic particle swarm optimization maximum power point tracker for photovoltaic system under partial shading condition," *IEEE Transactions on Industrial Electronics* 2013, vol. 60,8, pp. 3206-3195.

[91] Ishaque K, Salam Z, Amjad M, Mekhilef S, "An improved Particle Swarm Optimization (PSO)–based MPPT for PV with reduced steady-state oscillation," *IEEE Transactions on Power Electronics* 2012, vol. 27,8, pp. 3638-3627.

[92] Liu Y H, Huang S C, Huang J W, Liang W C, "A particle swarm optimization-based maximum power point tracking algorithm for PV systems operating under partially shaded conditions," *IEEE Transactions on Energy Conversion* 2012, vol. 27,4, pp. 1035-1027.

[93] Eltawil Mohamed A, Zhengming Zhao, "MPPT techniques for photovoltaic applications," *Renewable and Sustainable Energy Reviews* 2013, 25, pp. 813-793.

[94] Ishaque Kashif, Zainal Salam, "A review of maximum power point tracking techniques of PV system for uniform insolation and partial shading condition," *Renewable and Sustainable Energy Reviews* 2013, vol. 19, pp. 488-475.

[95] Subudhi Bidyadhar, Raseswari Pradhan, "A comparative study on maximum power point tracking techniques for photovoltaic power systems," *IEEE Transactions on Sustainable Energy* 2013, vol. 4,1, pp. 98-89.

[96] Reza Reisi, A Hassan Moradi M, Jamasb S, "Classification and comparison of maximum power point tracking techniques for photovoltaic system, A review," *Renewable and Sustainable Energy Reviews* 2013, vol. 19, pp. 443-433.

[97] Ahmed Jubaer, and Zainal Salam. "A Maximum Power Point Tracking (MPPT) for PV system using Cuckoo Search with partial shading capability." *Applied Energy* 119 (2014): 118-130.

[98] Yang X-S, Deb S. Multiobjective cuckoo search for design optimization. *Compute Oper Res* 2011.

[99] Socha Krzysztof, and Marco Dorigo. "Ant colony optimization for continuous domains." *European journal of operational research* 185, no. 3 (2008): 1155-1173.

[100] Liao Tianjun, Marco A Montes de Oca, Dogan Aydin, Thomas Stützle, and Marco Dorigo. "An incremental ant colony algorithm with local search for continuous optimization." In *Proceedings of the 13th annual conference on Genetic and evolutionary computation*, pp. 125-132. ACM, 2011.

[101] Chen Ling, Jie Shen, Ling Qin, and Hongjian Chen. "An improved ant colony algorithm in continuous optimization." *Journal of Systems Science and Systems Engineering* 12, no. 2 (2003): 224-235.

[102] Jiang, Lian Lian, Douglas L. Maskell, and Jagdish C. Patra. "A novel ant colony optimization-based maximum power point tracking for photovoltaic systems under partially shaded conditions." *Energy and Buildings* 58 (2013): 227-236.

[103] Parlak, Koray Şener, "PV array reconfiguration method under partial shading conditions," *International Journal of Electrical Power & Energy Systems* 2014, vol. 63, pp. 721-713.

[104] Chong B V P, L Zhang, "Controller design for integrated PV–converter modules under partial shading conditions," *Solar Energy* 2013, vol. 92, pp. 138-123.

[105] Lu F, Guo S, Walsh T M, Aberle A G, "Improved PV module performance under partial shading conditions," *Energy Procedia* 2013, vol. 33, pp. 255-248.

[106] Dolara A, Lazaroiu G C, Leva S, Manzolini G, "Experimental investigation of partial shading scenarios on PV (photovoltaic) modules," *Energy* 2013, vol. 55, pp. 475-466.

[107] Murtaza A, Chiaberge M, Spertino F, Boero D, De Giuseppe M, "A maximum power point tracking technique based on bypass diode mechanism for PV arrays under partial shading," *Energy and Buildings* 2014, vol. 73, pp. 25-13.

[108] *OutBack Power*, 17825 59th Ave. NE, Suite B Arlington, WA 98223 United States, http//www.outbackpower.com.

[109] "Genasun" LLC, 1035 Cambridge St. Suite 16B Cambridge, MA 02141 USA, http//genasun.com.
[110] "Blue Chip Energy Solution Pvt. Ltd.," E-57 Ground Floor Chattarpur Extn., New Delhi-110074. http//www.bluechipenergy.in.
[111] "Blue Sky Energy," HQ 2598 Fortune Way, Suite K Vista, CA 92081 USA. http//www.blueskyenergyinc.com.
[112] "Morningstar Corporation," 8 Pheasant Run, Newtown, PA 18940 USA, http//www.morningstarcorp.com.
[113] http//www.steca.com.

BIOGRAPHICAL SKETCHES

Deepak Verma

Affiliation: Department of Electrical & Electronics Engineering, Birla Institute of Technology Mesra, Jaipur Campus, Jaipur, RJ, India

Research and Professional Experience: 10 Years

Professional Appointments: Assistant Professor in Department of Electrical & Electronics Engineering, Birla Institute of Technology Mesra, Jaipur Campus, Jaipur, RJ, India

Publications from the Last 3 Years:

International Journals:

1. Nikhil Kumar, Savita Nema, Deepak Verma, "A State of art review on Conventional, Soft Computing & Hybrid Techniques for shading Mitigation in PV Applications" *International Transactions on Electrical Energy Systems* [Accepted for publication], 2020. [SCI and SCOPUS indexed]

2. Shilpa Kalambe, Sanjay Jain, Deepak Verma, "Enhanced Loadability and Inapt Locations" *International Journal of Recent Technology and Engineering*, Volume-8 Issue-4, November 2019. [SCOPUS indexed]
3. Deepak Verma, Nikhil Kumar, Shilpa Kalambe, "Tracking of Maximum Power Point in Solar PV (SPV) Systems using Perturb & Observe (PO) and Incremental Conductance (IC) Method" *International Journal of Innovative Technology and Exploring Engineering*, Volume-8, Issue-12, October 2019. [SCOPUS indexed]
4. Anand Kumar Singh, Nalin Harsh Vardhan, Deepak Verma, "Simulation and Experimentation of a Single Stage Boost Inverter." *International Journal of Engineering and Advanced Technology* (IJEAT), Vol-8 No-6, 2019: 1767-1774. [SCOPUS indexed]
5. Amarnath, R.K. Nema, Deepak Verma, "Modeling and Simulation of Solar Photovoltaic Application Based Multilevel Inverter with Reduced Count Topology for Stand-alone System." *Electrical & Computer Engineering: An International Journal (ECIJ)* 6 (2017):1-12
6. Das, Soubhagya K., Deepak Verma, Savita Nema, and R. K. Nema. "Shading mitigation techniques: State-of-the-art in photovoltaic applications." *Renewable and Sustainable Energy Reviews* 78 (2017): 369-390. [SCI and SCOPUS indexed]

International Conferences:

1. Keshav Gupta, Monika Sharma, Deepak Verma, "Design and fabrication of two stage 2 kVA inverter using push-pull topology" 10th *IEEE International Conference on Communication and Signal Processing* [ICCSP 2020], TN, India, 9-11th April 2020.
2. J. Majhi, S. Nema and Deepak Verma, "Simulation And Analysis Of Solar Based Water Pumping System For Hydraulic Actuated Load," 2019 *2nd International Conference on Power and*

Embedded Drive Control (ICPEDC), Chennai, India, 2019, pp. 333-341.
3. Deepak Verma, S Nema, R K Nema, "Implementation of Perturb and Observe Method of Maximum Power Point Tracking in SIMSCAPE/MATLAB" *IEEE International Conference on Intelligent Sustainable Systems, Palladam, Tamil Nadu*, Dec 2017.
4. Amarnath, R.K. Nema, Deepak Verma, "Harmonics Mitigation Of P&O MPPT Based Solar Powered Five-Level Diode-Clamped Multilevel Inverter" *IEEE International Conference on Innovations in Control, Communication and Information System 2017*.

Nikhil Kumar

Affiliation: Department of Electrical Engineering, Maulana Azad National Institute of Technology, Bhopal, MP, India

Research and Professional Experience: 4+ years

Professional Appointments: Research Scholar in Department of Electrical Engineering, Maulana Azad National Institute of Technology, Bhopal, MP, India

Savita Nema

Affiliation: Department of Electrical Engineering, Maulana Azad National Institute of Technology, Bhopal, MP, India

Research and Professional Experience: 20+ Year in Department of Electrical Engineering, Maulana Azad National Institute of Technology, Bhopal, MP, India

Professional Appointments: Professor, Department of Electrical Engineering, Maulana Azad National Institute of Technology, Bhopal, MP, India

In: Maximum Power Point Tracking ISBN: 978-1-53618-164-7
Editor: Maurice Hébert © 2020 Nova Science Publishers, Inc.

Chapter 2

MPPT CHARGE CONTROLLER FOR BATTERY CONNECTED PHOTOVOLTAIC POWER CONDITIONING UNIT

Joydip Jana[*], *Hiranmay Samanta,*
Konika Das Bhattacharya and Hiranmay Saha

Centre of Excellence for Green Energy and Sensor Systems,
Indian Institute of Engineering Science and Technology Shibpur,
Howrah, West Bengal, India

ABSTRACT

Photovoltaic (PV) systems include storage batteries when there is surplus power. This is used for providing electricity on demand. A suitable charge controller is needed for interfacing the solar PV module(s) with the battery bank. In this work, attention has been made to attribute more features to the controller which will enhance the efficiency and

[*] Corresponding Author's E mail: joydipjana02@gmail.com

controllability and most importantly will monitor the health of the battery being charged, ensuring a long batterylife.

Further, as the Maximum Power Point (MPP) of a solar PV array varies with solar irradiance and temperature, the developed controller has the facility to accurately track the MPP during static (fixed irradiance) and dynamic (rapidly changing irradiance) both weather conditions. This development focuses on the Battery charging from the PV modules at its dynamically changing MPP and at the same time follows a novel charging method which keeps checking on the condition of the battery health. An integrated software running from an embedded platform operating using the DSPIC microcontroller has been carried out. Both these aspects, of MPPT and efficient Battery charging were generally investigated separately earlier. The charge controller has been tested with Modules from 150W to 320W and has been found having minimum converter efficiency of 93.8%, and it can extract as much power as possible to charge the battery with an MPP tracking speed of 1 second and maximum MPP tracking efficiency of 99.9% with a fluctuation of 0.86% around the target MPP in static irradiation condition. And in dynamic irradiation conditions, the before mentioned performance parameters become 1.67 seconds, 98.5% and 0.50% respectively.

Keywords: MPPT, multi-stage battery charging, efficiency of charging, battery life extension, synchronous converter

1. INTRODUCTION

Battery Connected Photovoltaic (PV) systems include storage batteries for providing electricity on demand [1]. A suitable charge controller is needed for interfacing the solar PV module(s) with the battery bank which should protect the battery health [2, 3]. The primary job of these charge controllers is to charge the battery efficiently by following a charging method which will prevent the battery overcharge and over-discharge ensuring a long battery life.

Further, as the solar insolation and temperature has a considerable effect on Maximum Power Point (MPP) of the solar PV modules [4], the charge controllers for Battery Back-up Grid-Import PV system has to perform another task which requires a fine and fast tracking methodology with a deterministic approach. The challenge lies during conditions of minimal fluctuations of the PV operation around the MPP under both static (fixed irradiation) and dynamic (rapidly changing irradiation) weather conditions. During these conditions, the device should operate with precision, speed and efficiency. However, both these features of the battery charge controller for PV systems have been investigated separately in earlier works, the technical review of which is presented in an integrated fashion [4].

Various charging methods for batteries are reported in the literature: single stage, and multi-stage. The single-stage methods e.g., the constant-voltage (CV), the constant-current (CC), and the on-off method often present the problem of not fully charging the battery, and also not protected from premature aging of batteries [5]. It has been shown in [6] that, the efficiency of the multi-stage charging e.g., constant-current-constant-voltage (CCCV) is better than single-stage method no matter what type of battery it is. Multi-stage charge controllers generally have three charging stages some of which are CV and the rest are CC charging [7]. Some additional charging stages may be incorporated in the charge controller to enhance the life cycle of the battery and to enhance the performance of the battery over the charge and discharge periods.

For charging batteries using the PV module(s) a variety of MPPT techniques have been developed so far as reported in [8-16]. These techniques can be grouped into categories as (1) offline techniques which are dependent on PV cell models, (2) online techniques which are not specifically dependent on PV cell models and (3) hybrid techniques which are a combination of the two aforementioned techniques. But two primary issues associated with these techniques are "PV operating point fluctuation around the target MPP" and "speed of MPP tracking" [4].

This chapter first provides a review of the traditional battery charge controllers pointing out their limitations. A new MPPT charge controller has been developed that resolves limitations of the traditional controllers. The new controller can charge four types of batteries, namely, Flooded Lead acid battery, Valve regulated lead acid (VRLA) or sealed lead acid battery, VRLA Gel battery and VRLA Absorbent glass mat (AGM) battery. It on one hand brings the battery to a full charge efficiently by five-stage charging ensuring long battery lifetime and on the other hand tracks the MPP very fast and efficiently with small PV operating point fluctuation around the MPP under both static and dynamic weather conditions by using a variable step size perturb and observe (P&O) MPPT technique. The efficiency of the converter used in the controller has been enhanced through some special means. Additionally the need to provide a very wide input voltage range in the charge controller has also been addressed in this work [17].

Various experimental tests have been performed in order to verify the performance of the new charge controller.

2. REVIEW OF MPPT CHARGE CONTROLLERS

2.1. Multi-Stage Charging

Various charge controllers for charging batteries have been introducedin the past in keeping with the substantial development in battery technologies from flooded lead-acid to VRLA as reported in the literature [18-20]. Among them, the CCCV charge controller has been the most commonly used [21]. They generally have three charging stages namely, bulk, absorption, and float respectively. If constraints are put on the states of the battery such as the voltage of the battery, the state-of-charge of the battery, the charging current, and the temperature of the battery which are typically given by the manufacturer of the battery, the life degradation of the battery can be minimized [24-27]. Such constraints are expected to be followed in the three-stage CCCV charge controllers [28]. In the bulk charging stage, the battery is generally charged with a charging current of C/10 Amp (C = capacity of the battery). Batteries can be charged faster with high set points of the charging current and/or the charging voltage [22, 23]. But the charging faster beyond recommendation level puts a negative effect on the various aging factors of the battery which results in reduced battery life by sulfation in the negative electrode, water loss, and grid corrosion [21]. Another issue is thatif the battery condition is not good enough to accept a high amount of current during bulk charging then low current should be applied before bulk stage only to make the battery healthy to accept high current charging.

The last stage in a traditional three-stage charge controller is called the float stage where the battery charging voltage is lowered than that of the previous stage and constant voltage is applied. If left in an extended float state, the battery faces the threat of acid sulfate *stratification*. Separation of the acid and the water of the electrolyte makes acid deposition at the bottom of the battery surface leading to Stratification, ultimately reducing the performance and life of the battery. Thus an additional stage of charging is required to maintain a balance between battery overcharging and undercharging and to prevent battery stratification.

2.2. MPPT Techniques

Continuous efforts have been made by the researchers on the design and development of the MPPT techniques for a PV system to attain improved performance parameters. The objectives are: increasing the speed of MPP tracking even when there are small fluctuations in the PV operating point around the MPP during constant as well as rapidly varying weather conditions. In this section, a review of different MPPT techniques found in literature has been presented to depict their performance characteristics and limitations.

An MPPT charge controller as described in [29] is a type of battery charge controller that ideally brings the operating point of a PV module to its MPP [30]. It is widely used in PV systems to increase their MPP tracking efficiency [31] and to extract the maximum possible instantaneous power from the PV module [32]. As there are a variety of MPPT techniques available in the literature, it is quite difficult to conclude a particular technique to be better. The selection can be made based on the application, DC-DC converter used, MPP tracking time and MPP tracking efficiency. Some of the extensively used MPPT techniques based on perturb and observe (P&O) and incremental conductance (INC) methods that have been proposed and implemented in hardware are discussed in detail in [33-44].

In Table 2.1, a few MPPT techniques which have been implemented in hardware are listed and compared based on the DC-DC converter used, the controller used and MPP tracking time.

Some artificial intelligence MPPT techniques such as the artificial neural network (ANN), fuzzy logic control (FLC), etc. and MPPT techniques iterative in nature like genetic algorithm (GA) are discussed in [48-50]. Two primary parameters associated with these MPPT techniques are "PV operating point fluctuation around the target MPP" and "speed of MPP tracking" and there is a tradeoff between these two parameters which need to be considered. In Table 2.2, the MPP tracking time and the operating point fluctuation around the MPP for different MPPT techniques have been presented.

Table 2.1. MPPT techniques based on P&O and INC method

MPPT technology	Converter used	Controller	MPP tracking time
FLC based P&O [45]	Buck converter	DSP TMS320F28335	1.5s
Power increment INC [46]	Flyback converter	dsPIC33FJ06GS202	5s
Modified variable step-size INC [47]	Buck converter	PIC 18F4520	2s

Table 2.2. MPP tracking time and the PV operating point fluctuation around target MPP for different MPPT techniques

MPPT technique	MPP tracking time	Operating point fluctuation around the target MPP
stand-alone FLC [51]	27 ms	2%
combination of ANN and P&O [52]	2 ms	nil
combination of ANN and I&C [53]	40 ms	-
combination of ANN and GA [54]	-	1.45%
GA optimized FLC [55]	715 ms	0.02%
HPNN coupled with FLC [56]	10-40 ms	0.002%
standalone FLC [57]	150 ms	-
FLC integrated with HC [58]	70 ms	-
adaptive fuzzy logic control [59]	3 ms	-
second-order sliding mode-based MPPT [60]	0.2 s	-

It has to be noted that, amongst all the MPPT techniques based on artificial intelligence as mentioned in Table 2.2, none are implemented in hardware, and hence their practical feasibility could not be established. It is pointless to use a more complex or more expensive MPPT technique if a simpler and less expensive one leads to similar results. This may be the reason why some computationally intensive techniques are not implemented in hardware.

3. Description of the Developed MPPT Based Charge Controller

An improved maximum point tracking (MPPT) based battery charge controller has been designed and developed for operation in Battery Back-up Grid-Import PVPCU (Power Conditioning Unit). The developed charge control algorithm regulates the duty cycle of the switching devices of the DC-DC converter of the charge controller following a modified perturb and observe (P&O) technique. At the same time, it provides precise control on the battery charging voltage and current according to the five charging stages to ensure a maximum charging rate and simultaneously protecting the battery health. Both these features, which have been generally investigated separately earlier, are treated in an integrated fashion in this work.

The work analyses in detail about the DC-DC converter power loss as well as describes the development of the MPPT technique and battery charging method used in the charge controller. A prototype of 1500W power delivering capacity has been designed, fabricated and tested. The problems of premature aging, slow MPP tracking speed, large PV operating point fluctuations around the target MPP, low MPP tracking efficiency (for static and dynamic weather conditions) have been addressed with proper care in this work.. The converter efficiency has also been improved. Besides these some aspects such as dependence on the battery type, incompatibility with the large range of input PV voltage, temperature compensation, etc. have also been addressed satisfactorily. Moreover, two additional stages namely soft charging and equalization charging have been applied in the developed charge controller along with the three existing stages of the traditional charge controller to achieve the balance between battery overcharging and undercharging and to prevent battery stratification.

3.1. Improved DC-DC Converter

A prototype of the charge controller of 1500W power transferring capacity has been developed (Figure 1). Asynchronous DC-DC buck converter has been used instead of using the conventional non-synchronous DC-DC buck converter.

A power diode has been used in the developed converter for reverse polarity protection at the battery terminals of the charge controller and backflow (current flow from battery side to PV module side) protection. A solid-state relay has been used in parallel with this power diode as shown in Figure 1, which bypasses the power diode above a certain delivering power to minimize the power loss across the diode as the voltage drop of the diode is high. These design considerations have been taken to enhance the power conversion efficiency of the charge controller.

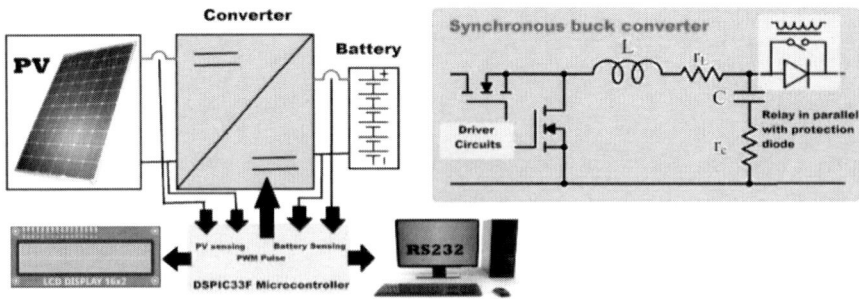

Figure 1. Schematic diagram of the synchronous buck converter in the developed charge controller.

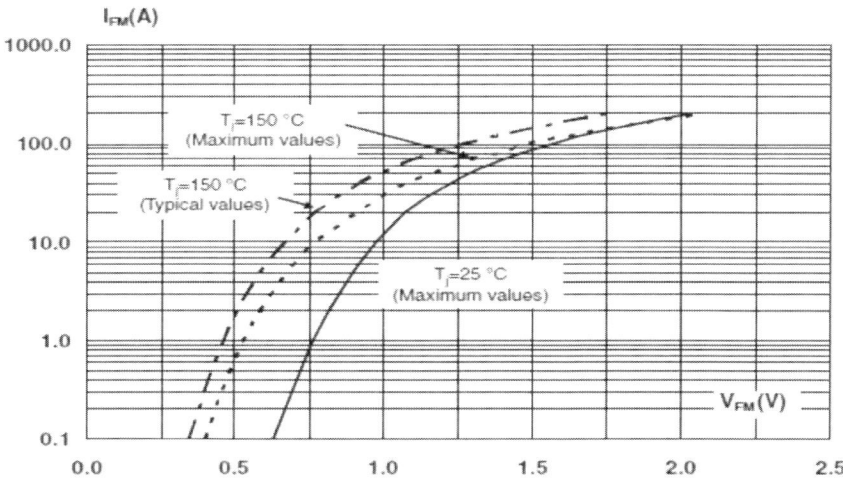

Figure 2. Forwarding current vs. the forward voltage drop of the power diode (STTH60W02C) used in the DC-DC converter.

From Figure 2, it can be seen that for delivering 30Amp (full capacity) current to the battery for the designed synchronous DC-DC converter, the diode voltage drop will be 1.25V (approximately), resulting in a power loss of 37.5W. If the relay bypasses the power diode then the power loss in the relay will be 1.8W, resulting in a reduction of 2% power loss. Above a certain output loading of the converter, the power diode is bypassed by the relay, and this mechanism is controlled by the microcontroller used in the battery charge controller.

The hardware of the prototype is composed of four main sections, namely the

- power section consisting of power switching devices with their driver circuits,
- the control section,
- the sensing section, and
- the display.

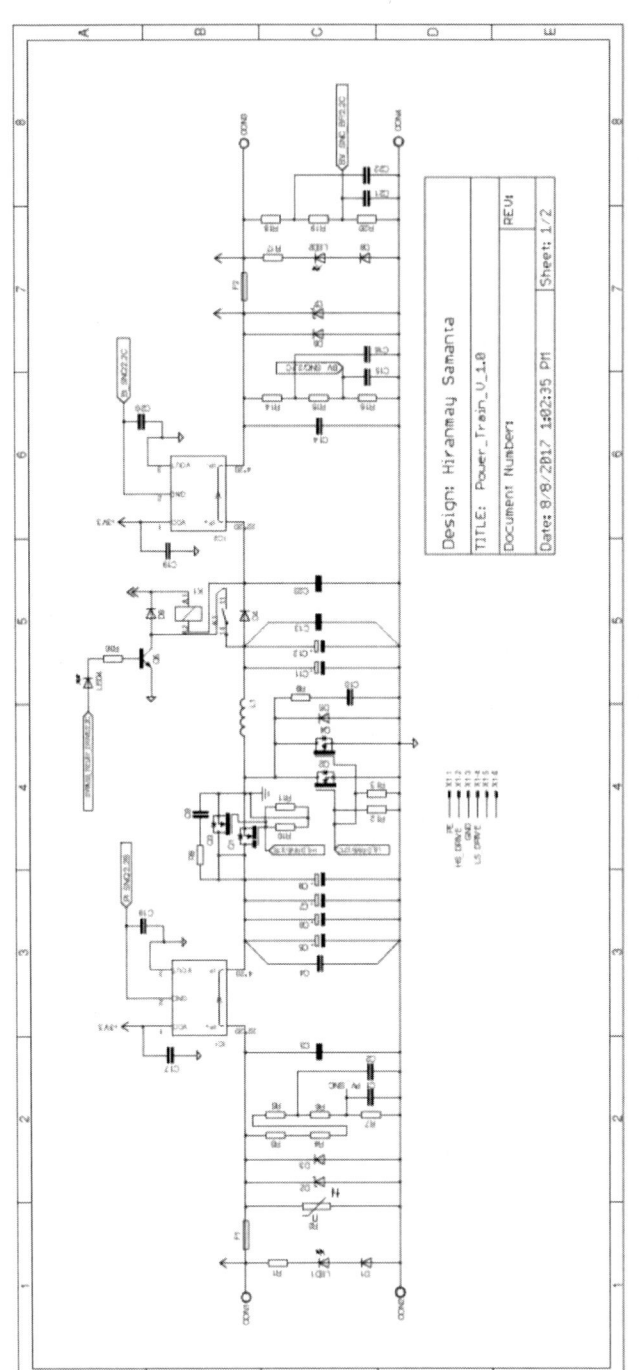

Figure 3. Circuit diagram of the developed synchronous buck DC-DC converter.

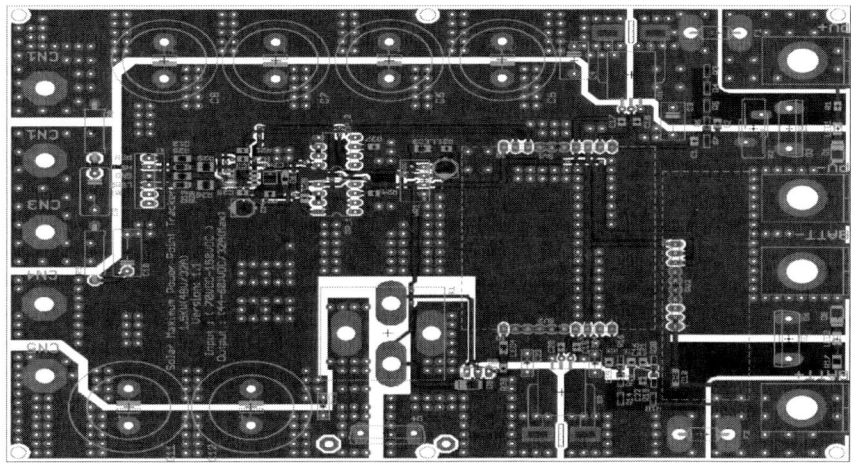

Figure 4. PCB layout of the developed synchronous buck DC-DC converter.

The circuit diagram and PCB layout of the hardware are shown in Figure 3 and in Figure 4 respectively. The control section contains a DSPIC33F microcontroller. The main function of this section is to generate the PWM pulses with variable duty following the MPPT technique, which drives switching devices of the DC-DC converter using the driver circuits. The sensing section comprises a negative temperature co-efficient thermocouple for temperature sensing, resistive voltage divider based sensors for PV and battery voltage measurement and Hall based current sensors for PV and battery current measurements. The hardware contains a 16*2 LCD to display various charging parameters.

Enhancement of efficiency of the converter is one of the primary design approach; therefore the power loss of the system needs to be determined. The major losses incurred in the components of the synchronous buck converter for the prototype development has been represented through equation1to 8 [57]. However to understand better the turn-on and turn-off phenomena of the MOSFET in DC-DC or DC-AC power converters, the clamped inductive switching mode of operation of the MOSFET embedded in a simple non-isolated DC-DC boost converter has not been presented in detail in this article.

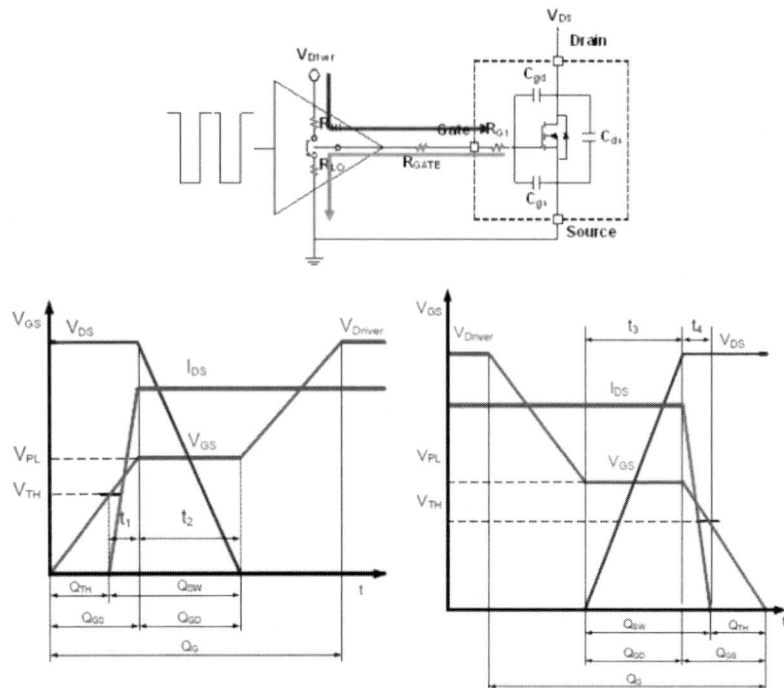

Figure 5. (a) MOSFET driver turns on and off path, (b) MOSFET driver turn-on waveforms, (c) MOSFET driver turn-off waveforms.

The Figure 5 shows the voltage waveforms during the MOSFET driver turn on and off situations whereas Figure 6 representthe current paths when the high-side and low-side of the MOSFET conduct respectively.

The high-side MOSFET's switching on loss is given in equation (1) [61],

$$P_{hs\_on} = f_{SW} \times V_{DS} \times I_{DS} \times \frac{t_1 + t_2}{2} = f_{SW} \times V_{in} \times I_{out} \times \frac{T_{hs\_on}}{2} \qquad (1)$$

where, f_{SW} = switching frequency, V_{DS} = Drain-source voltage, I_{DS} = Drain current, t_1 = Drain current rise time, t_2 = Drain-source voltage fall time of high-side MOSFET.

$$\text{Where, } \frac{T_{hs\_on}}{2} = \frac{Q_{SW}}{I_{G\_on}} \text{ and } I_{G\_on} = \frac{V_{driver} - V_{PL}}{R_{HI} + R_G + R_{G1}} \qquad (2)$$

where, V_{driver} = Voltage applied at the gate of the device, V_{PL} = Miller plateau level of gate-source voltage, R_G = Gate resistance.

The conduction loss of high-side MOSFET is determined by the on-resistances of the device and the device RMS current which is as follows:

$$P_{con\_hs} = I_{RMS\_HG}^2 \times R_{DS\_on\_hs} \tag{3}$$

where, I_{RMS_HG} = RMS value of the conducting current, $R_{DS\_on\_hs}$ = Resistance of the high-side MOSFET

$$\text{where, } I_{RMS\_HS} = \sqrt{D \times (I_{out}^2 + \frac{I_{ripple}^2}{12})} \tag{4}$$

where, I_{out} = output current of the converter, I_{ripple} = Ripple in the output current.

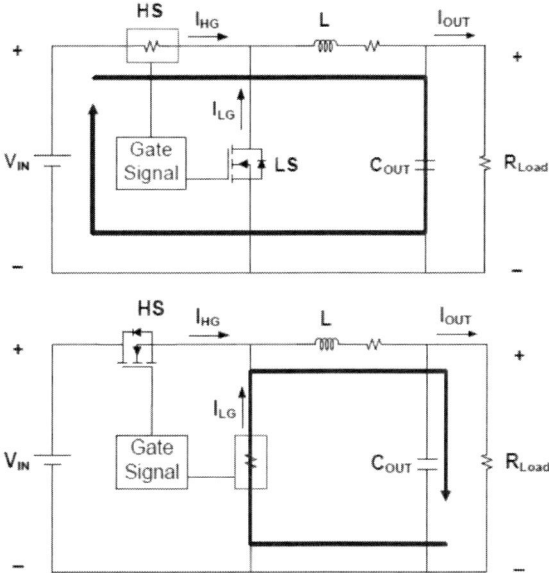

Figure 6. (a) Equivalent circuit when high-side MOSFET conducts, (b) Equivalent circuit when low-side MOSFET conducts.

The conduction loss of low-side switching device is as follows,

$$P_{con\_ls} = I^2_{RMS\_LG} \times R_{DS\_on\_ls} \qquad (5)$$

where, I_{RMS_LG} = RMS value of the conducting current, $R_{DS\_on\_ls}$ = Resistance of the low-side MOSFET

$$\text{where, } I_{RMS\_LG} = \sqrt{(1-D) \times (I^2_{out} + \frac{I^2_{ripple}}{12})} \qquad (6)$$

where D = duty ratio of the on-time of the high-side and low-side MOSFET.

The dead-time loss induced by low-side device body diode conduction during dead-times is as follows,

$$P_{deadtime} = V_{SD} \times [\left(I_{out} - \frac{I_{ripple}}{2}\right) \times T_{D2} + (I_{out} + \frac{I_{ripple}}{2}) \times T_{D1}] \times f_{SW} \qquad (7)$$

and the low-side device reverse recovery charge loss is as follows,

$$P_{rr} = Q_{rr} \times V_{DD} \times f_{SW} = Q_{rr} \times V_{in} \times f_{SW} \qquad (8)$$

where, Q_{rr} = Reverse recovery charge, V_{in} = Input voltage of the converter.

3.2. Improvement in Maximum Power Point Tracking Technique

The benefit of the P&OMPPT technique is that there is a fixed step size of PV voltage perturbation and it is easy to implement. However, the limitations of this technique are low tracking speed and high PV operating point fluctuations around the target MPP [62]. The accurate step size of PV voltage perturbation is important to provide good performance under fast-changing weather conditions [63-65]. To solve the fore mentioned limitations, some modified MPPT techniques with variable step size had been proposed in the literature and simulated [66–69], but none are implemented in hardware, and hence their practical feasibility could not be established. The developed charge controller employs a variable step size P&O MPPT technique to improve the tracking speed of MPP and to lower the PV operating point fluctuations around MPP. An important reason for selecting the P&O technique is that it is not PV module size-dependent and the charge controller can have a very wide input voltage range and thereafter be compatible with a wide range of PV modules. The performance of the MPPT technique has been experimentally validated which has been discussed later in this chapter.

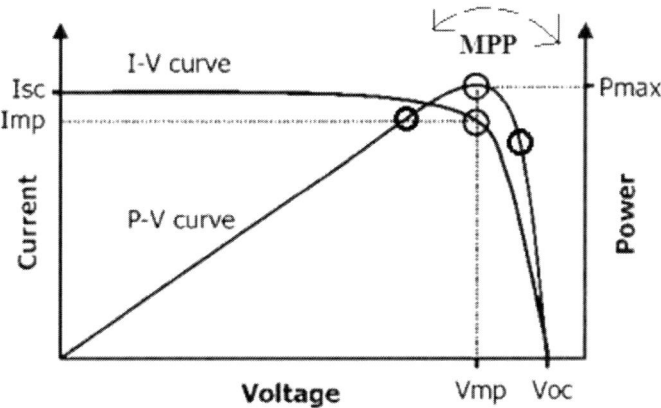

Figure 7. The fluctuation of the PV operating point around MPP with a big step size of PV voltage perturbation.

The duty cycle of the switching device in the converter is set according to the requirement of the PV voltage perturbation step size and it varies to balance between two conflicting parameters, namely the speed of MPP tracking and the fluctuation of the PV operating point around the MPP. When the charge controller starts the MPP tracking process, it executes a dynamic pilot loop program that traverses the I-V characteristics curve of the PV module from the open circuit point towards the MPP. In a dynamic MPPT loop, PV voltage is perturbed with a big step size to catch the MPP fast as shown in Figure 7. The dynamic MPPT loop also is applied in that situation when there is a big change in PV power generation or the irradiation is changing very fast.

Table 3.1. The change of step size of PV voltage perturbation according to the change in PV power

Change of PV power	A step size of PV voltage perturbation
ΔPV_Power_1	$\Delta Step\_Size_1$
$\Delta PV\_Power_2 > \Delta PV\_Power_1$	$\Delta Step\_Size_2 > \Delta Step\_Size_1$

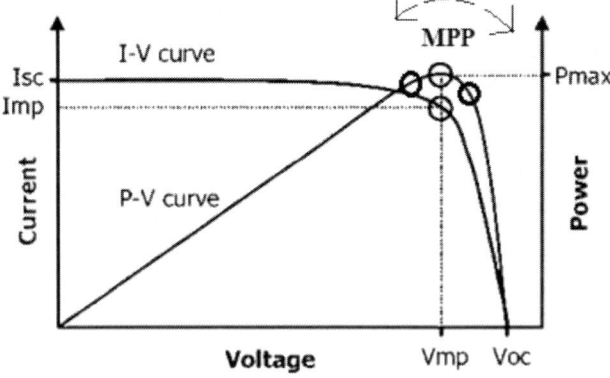

Figure 8. The fluctuation of the PV operating point around MPP with a small step size of PV voltage perturbation.

In a dynamic MPPT loop, the step size of PV voltage perturbation is high leading to a fluctuation of the PV operating point around the MPP. Once the charge controller catches the MPP, then a static MPPT loop is applied where the step size of the PV voltage perturbation becomes small to attain low fluctuation of the PV operating point around the MPP which has been shown in Figure 8. And when there is no change in irradiation or there is a slow change in irradiation, then also the static MPPT loop is applied. This is the main idea of performance optimization between the tracking speed of MPP and the PV operating point fluctuations around MPP. Now until there is a big power change due to sudden fast change of solar irradiation the program loop stays in this static MPPT and it stays here very stable with high static MPP tracking efficiency due to the small step size of the PV voltage perturbation. The change of the step size of PV voltage perturbation according to the change in PV power has been presented in Table 3.1. The static and dynamic loops swing between each other when there is a change of power equal to this set value. For example, if the charge controller is delivering power of 500W to the battery, then the set-point for power change is 0.5W i.e., if there is a power change of more than0.5W then it goes from static loop to dynamic loop to catch the new MPP quickly by increasing the step size of PV voltage perturbation.

3.3. Improvement in Multi-Stage Charging

The additional stages incorporated in the newly developed charge controller with three existing stages conventionally used in traditional charge controllers are soft charging and equalization. The set voltage in every charging stage for each of the batteries has been prepared and incorporated in the microcontroller program of the new charge controller. The schematic diagram of the developed charge controller with the flowchart of the battery charging method and the timing diagram of the charging voltage and charging current in the developed charge controller has been shown in Figure 9(a) and Figure 9(b) respectively. In the case of VRLA Gel batteries and VRLA AGM batteries, the equalization stage has been neglected.

The developed charge controller detects battery state-of-charge by detecting battery voltage before charging. After reading the battery voltage the charge controller determines which stage to properly charge at. In Table 3.2, the open-circuit voltage (OCV) for different state-of-charge (SOC) for different types of batteries has been shown. In Table 3.3, the type of charging whether it is constant current charging or constant voltage charging and the state of the charge for each charging stage has been shown.

Table 3.2. The open-circuit voltage (OCV) for different state-of-charge (SOC) for different types of batteries

Battery type	Open circuit voltage (OCV) @20% state-of-charge (SOC)	Open circuit voltage (OCV) @50% state-of-charge (SOC)	Open circuit voltage (OCV) @60% state-of-charge (SOC)	Open circuit voltage (OCV) @70% state-of-charge (SOC)	Open circuit voltage (OCV) @90% state-of-charge (SOC)
Flooded lead-acid battery	-	49.76V	50.08V	50.4V	51.08V
Sealed lead-acid battery or VRLA	48.4V	49.76V	50.08V	50.4V	51.08V
VRLA Gel battery	48V	49.6V	50.0V	50.4V	51.96V
VRLA AGM battery	48V	49.6V	50.0V	50.4V	51.96V

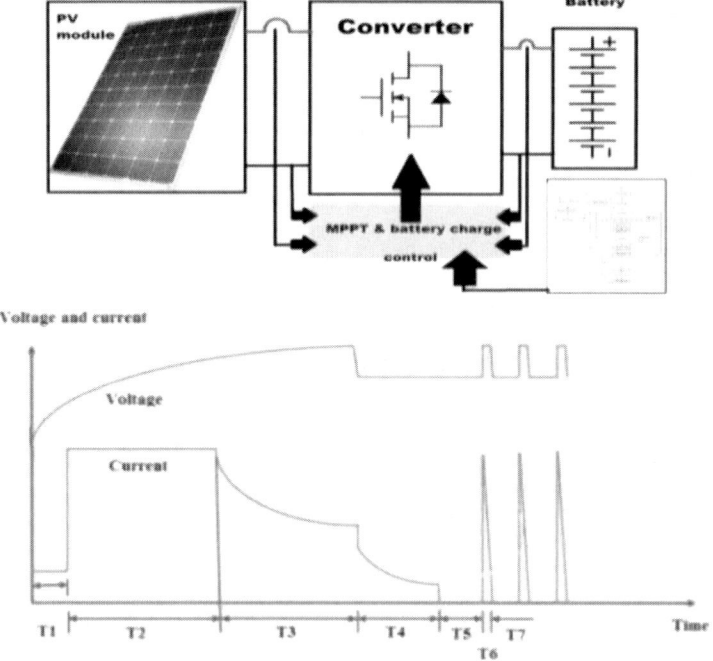

Figure 9. (a) Schematic diagram of the developed charge controller with the flowchart of the battery charging method, (b) Timing diagram of the charging voltage and charging current in the developed charge controller.

Table 3.3. The type of charging and the state of the charge for each charging stage

Charging stage	Type of charging	State-of-charge
Soft(T1)	Constant current charging with C(capacity)/100 current	Up to 20% state-of-charge (V_{Soft})
The bulk (T2)	Constant current charging with C(capacity)/10 current	20%-80% state-of-charge (V_{Bulk})
Absorption(T3)	Constant voltage charging	80%-100% state-of-charge ($V_{Absorption}$)
Float(T4)	Constant voltage charging	V_{float}
Equalization (T5)	Re-initiating of charging	-

Table 3.4. The voltage level of different types of battery in different charging stages

Charging stage	Battery type			
	Flooded lead-acid battery	Sealed lead-acid battery or VRLA	VRLA Gel battery	VRLA AGM battery
Soft(T1)	54V	53.6V	51.2V	51.2V
The bulk (T2)	59.2V	58.8V	56.4V	56.4V
Absorption(T3)	60.0V	59.6V	57.6V	57.6V
Float (T4)	54V – 55.2V	52.8V – 53.6V	54V – 55.2V	54.4V – 55.2V
Equalization(T5)	Charge voltage on point – 50.4V Charge voltage off point – 54V	Charge voltage on point – 50.4V Charge voltage off point – 52.8V	Charge voltage on point – 51V Charge voltage off point – 54V	Charge voltage on point – 51V Charge voltage off point – 54.4V

Table 3.5. Temperature compensation for different types of batteries (LUT 2)

Battery type	Temperature compensation
Flooded lead-acid battery	For average operating temperatures below this range (colder than) or above this range the maximum voltage set point should be compensated with an increase or with a decrease at a rate of 0.063 Volts Per Cell (1.512V Volts for a 48V battery) for every 10°C.
Sealed lead-acid battery or VRLA	1.512V Volts for a 48V battery for every 10°C.
VRLA Gel battery	1.2V Volts for a 48V battery for every 10°C.
VRLA AGM battery	0.432V Volts for a 48V battery for every 10°C.

In Table 3.4, the voltage level or state-of-charge in different charging stages has been shown for four types of batteries. The charge controller slightly changes the voltage set points for every stage according to the change of temperature and in this way the temperature is compensated. A look-up table (LUT 2) has been prepared and incorporated in the microcontroller program of the charge controller to determine the PV voltage set points with an increase/decrease rate of voltage for temperature change which has been shown in Table 3.5 for different types of batteries. The first charging stage in the developed charge controller is soft charging which is constant current charging. This stage has been applied only when the battery state-of-charge is below 20% or the battery voltage is lower than V_{Soft}. As in the bulk charging stage, the battery is charged with a high current and if the battery state-of-charge is lower than 20% then the battery is not capable to accept this high current. In this stage, if the PV module voltage is greater than the minimum input operating voltage of the charge controller then the PV module charges the battery with C/100 (C = Capacity of the battery in AH) current up to 20% state-of-charge. The next three charging stages named bulk, absorption, and float are the same as those of traditional charge controllers. As shown in Figure 9(b), T1, T2, T3, T4 signifies the soft, bulk, absorption, and float charging stages respectively and T5 and T6 denote the charge-on period and charge-off period during equalization stage respectively which has been discussed next.

The last stage is the equalization charging stage. In this stage, the developed charge controller monitors if the battery has remained in the float stage for a specified length of time or if the battery voltage drops below a minimum level. The charge controller then enters this stage by automatically initiating a new round of charging, correcting the undercharge condition and stimulating the mixing of the electrolyte solution. Specifically, the equalization stage of the developed charge controller automatically repeats the charge cycle every seven days or when the battery voltage drops below a determined voltage as shown in Figure 9. By using this technique, a balance is achieved between overcharging and undercharging and stratification is prevented. The construction of AGM and Gel batteries eliminates any stratification, thus this stage has not been applied in these batteries. After the equalization stage, the charge controller goes to the bulk charging stage and repeats the whole charging cycle.

4. EVALUATION OF THE DEVELOPED CHARGE CONTROLLER

Measuring the efficiency of MPPT techniques had not been standardized until the European Standard EN50530 was published at the end of May 2010 [70]. It specifies how to test the efficiency of MPPT techniques both in static (fixed solar irradiance) and dynamic conditions (changing solar irradiance).

An experimental setup has been prepared as shown in Figure 12 (b) to test the developed charge controller (shown in Figure 12 (a)), and to find out on its MPPT performance parameters. Two modes of operation have been arranged and examined. The tests have been performedbased on the standard EN50530. The first mode, when the charge controller has been operated with a fixed irradiation condition to find out its static MPP tracking efficiency. The second mode, when it has been operated with varying irradiation conditions to find out its dynamic MPP tracking efficiency. A PV simulator manufactured by Elgar TerraSAShas been used instead of the original PV module for charging a battery bank of 48V nominal voltage to test the developed charge controller within the laboratory.

The Elgar TerraSAS PV Simulator which is shown in Figure 10, is a DC power source which provides a programmable means of simulating the behavioral characteristic of a PV module or an array of PV modules. The instrument provides a turn-key approach to test the MPPT characteristics of the PV power conditioning units. The hardware of the PV simulator is controlled by a software program running on a personal computer that communicates directly to the PV simulator using an RS232 communication protocol, which functions as an IV characteristics curve generator. The

Graphical User Interface (GUI) of the software provides all of the user controls to the PV simulator.

The GUI of the software, which is shown in Figure 11, allows configuring the parameters of a PV module or an array of modules like the open-circuit voltage, V_{oc}, short-circuit current, I_{sc}, the voltage at the maximum power point, V_{mp} and current at the maximum power point, I_{mp} at 25°C and $1000W/m^2$, so that the resulting IV characteristics curve is calculated and generated according to a standard solar cell model. Changes to these parameters will permit the shape of the IV characteristics curve to be adjusted to any fill factor between 0.5 and 1. Once an IV characteristic curve has been generated, changes to the irradiation or temperature can be changed on the go so that the behavior of a PV power conditioning unit can be tested under naturalistic weather conditions for cloud shadowing and PV module temperature rise.

Figure 10. TerraSAS PV Simulator.

Different types of PV modules (Operating temperature: -40°C to +85°C, Temperature Co-efficient for Voc/°C (β): -0.34%, Temperature Co-efficient for Isc/°C (α): 0.05%, Temperature Co-efficient for power/°C (γ): -0.43%)can be used to charge batteries using this new charge controller and the charge controller has been tested with these PV module parameters settings in the PV simulator which have been shown in Table 4.1. For voltage and current measurement of the PV module and battery, a power meter manufactured by Yokogawa (model: WT330) has been used in this experimental set-up. The gate signals (captured in an oscilloscope) applied to the high-side and low-side MOSFETs have been shown in Figure 13.

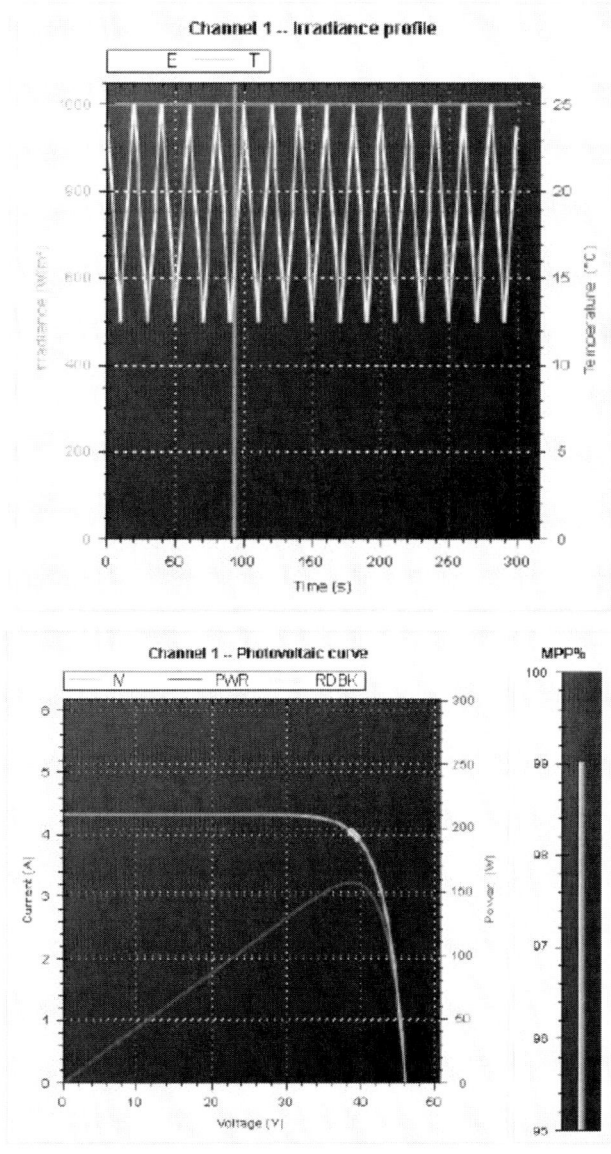

Figure 11. The GUI of the software of the TerraSAS PV Simulator.

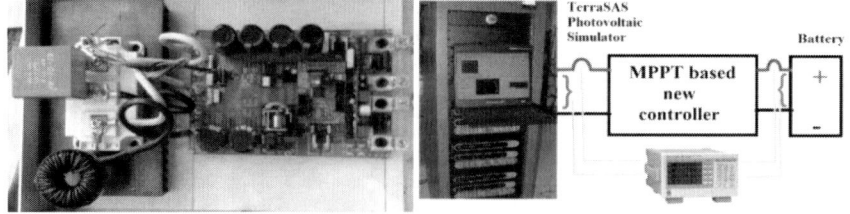

Figure 12. (a) The developed charge controller, (b) Experimental set-up.

Table 4.1. Parameters of the PV modules can be used for this charge controller

Parameter	Value							
Peak power	150W	205W	250W	300W	305W	310W	315W	320W
The voltage at MPP, Current at MPP	36.22V, 4.15A	36.18V, 5.67A	36.18V, 6.92A	36.14V, 8.31A	36.30V, 8.41A	36.40V, 8.52A	36.49V, 8.64A	36.56V, 8.76A
Open circuit voltage, Short circuit current	44.46V, 4.37A	44.41V, 5.97A	44.46V, 7.28A	44.46V, 8.74A	44.60V, 8.82A	44.71V, 8.92A	44.86V, 9.01A	45.01V, 9.11A
Module efficiency (%)	15.21%	14.09%	15.53%	15.63%	15.89%	16.15%	16.41%	16.67%

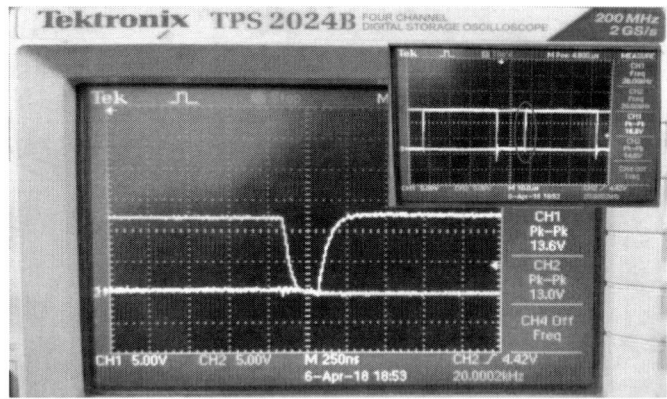

Figure 13. Gate signals of the high-side and low-side device captured an oscilloscope.

4.1. Static MPPT Performance Test

In this mode, the irradiation has been fixed at 120 W/m^2 and the temperature has been fixed at 25°C in the PV simulator. Figure 14, 15 and

16 show the curve of the PV voltage, PV current and PV power concerning time indicated in blue together with the ideal curve indicated in black.

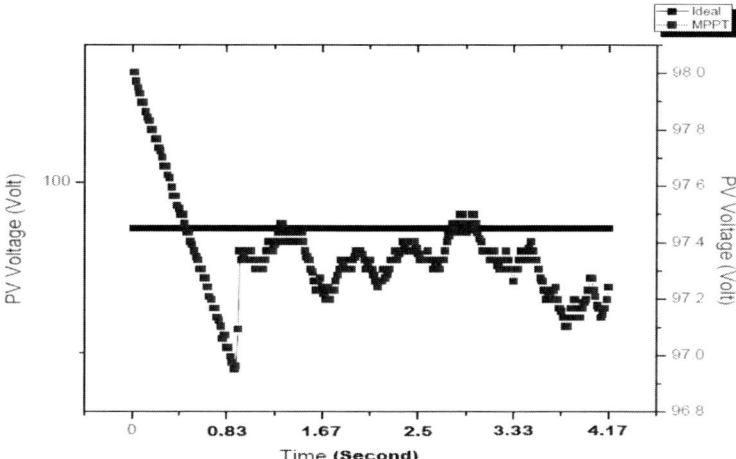

Figure 14. The output voltage of the PV with respect to time while charging in static irradiation condition.

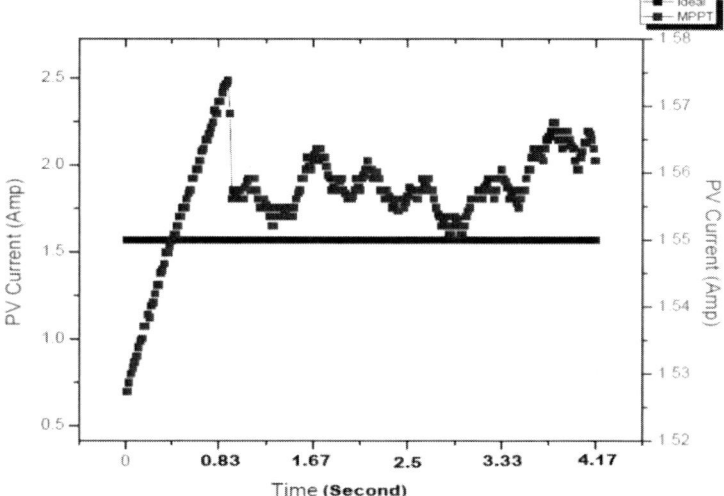

Figure 15. The output current of the PV with respect to time while charging in static irradiation condition.

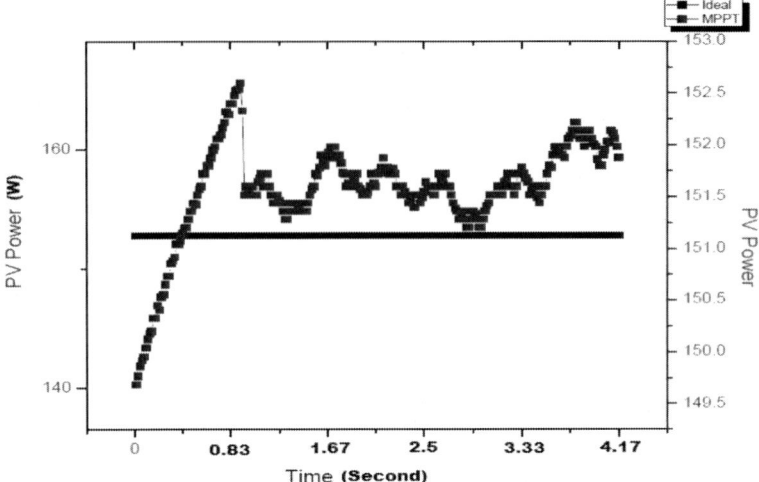

Figure 16. The output power of the PV with respect to time while charging in static irradiation condition.

The power meter used in this experiment can capture sixty numbers of measurements (data points) in a second for each parameter. According to Figure 16, it takes 1 second for the developed charge controller to reach the MPP and there a fluctuation of 0.86% around the target MPP has been achieved.

4.2. Dynamic MPPT Performance Test

In this mode, the irradiation has been varied in the PV simulator according to the dynamic irradiation profile shown in Figure 17. According to this dynamic profile, the irradiation has been varied from 100 W/m^2 to 1000 W/m^2 and sometimes the change of irradiation is so fast that it has been found very close to step jump of irradiation change.

To find out the dynamic MPP tracking efficiency of the new charge controller, some portions of the full dynamic response has been zoomed in, which have been shown in Figure 18, Figure 19 and Figure 20. The Figure 18 demonstrates that when the irradiation changes from 257 W/m^2 to 255 W/m^2, then the charge controller takes only 1.67 seconds to find the new MPP and when it catches the new MPP the fluctuation has been found 0.5% around the new MPP.

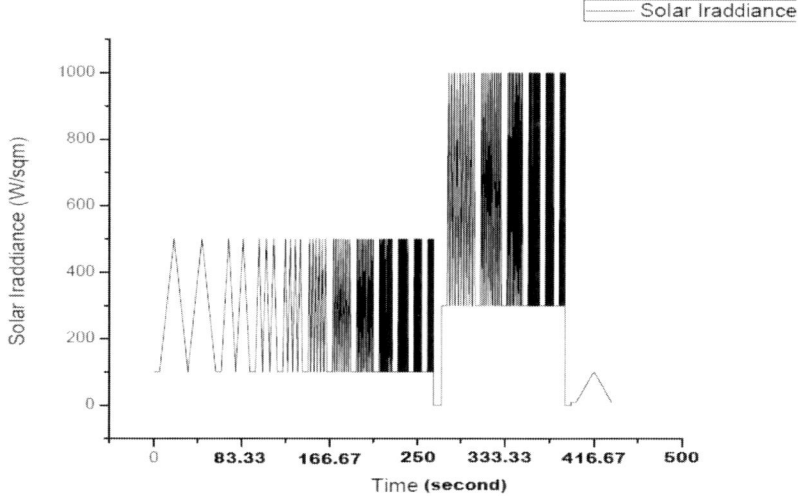

Figure 17. Dynamic irradiation profile used in the experiment.

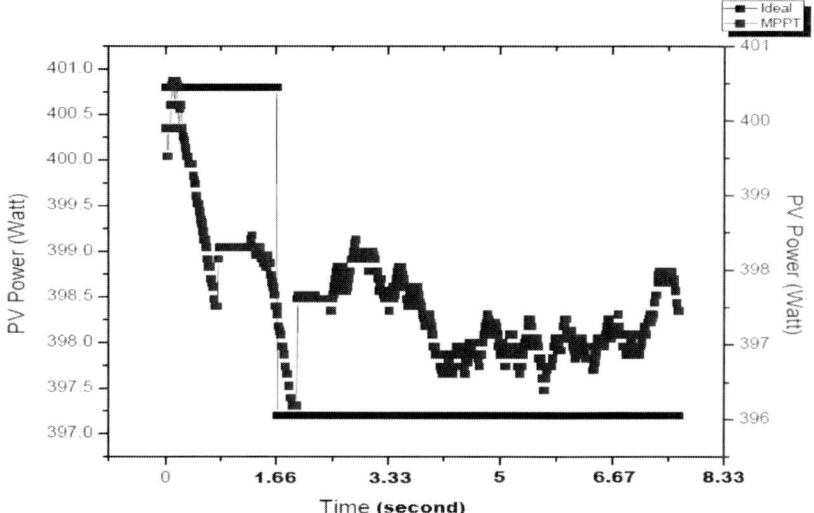

Figure 18. PV power in the dynamic condition (when the irradiation changes from 257 W/m^2 to 255 W/m^2.

In Figure 19, the PV current and MPP tracking efficiency according to the dynamic response of the charge controller have been shown where the irradiation has a nature of ramp-up from 100 W/m^2 to 500 W/m^2 first and then a ramp down from 500 W/m^2 to 100 W/m^2. It has been seen from the figure that the shape of changing the PV current follows the shape of the irradiation.

In Figure 20, the MPP tracking efficiency has been shown where the irradiation has a nature of ramp down from 500 to 100 W/m^2. The figure shows that the minimum dynamic MPP tracking efficiency i.e., MPP tracking efficiency in varying irradiation conditions has been found 96.5% and the maximum dynamic MPP tracking efficiency has been found 98.5% during the whole MPP tracking process according to the irradiation change.

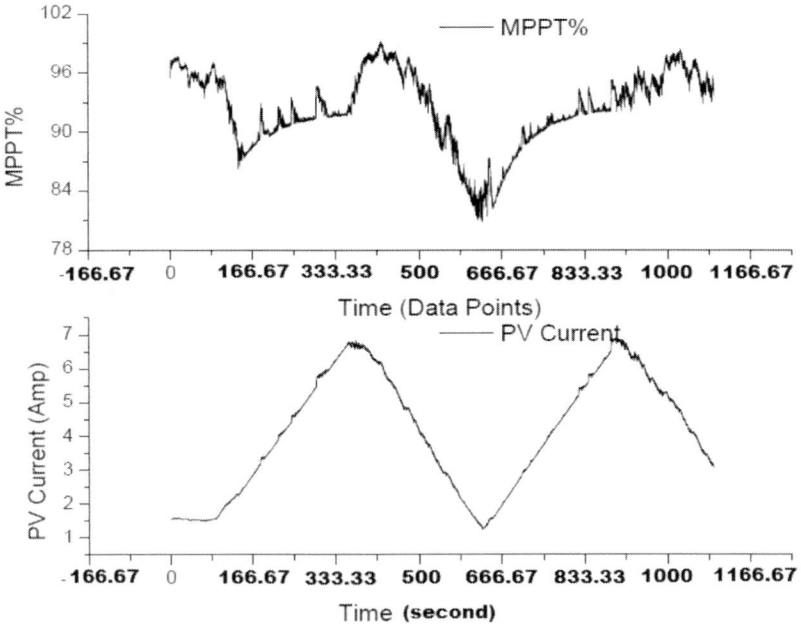

Figure 19. MPP tracking efficiency and PV current in dynamic condition (when the irradiation changes from 100 W/m^2 to 500 W/m^2).

4.3. Converter Efficiency Test

To find out the power conversion efficiency of the developed charge controller for a minimum load condition to full load condition, three sets of operation have been arranged and examined. The first set, when the charge controller has been operated with a PV configuration of V_{oc} = 85V in the TerraSAS PV simulator.

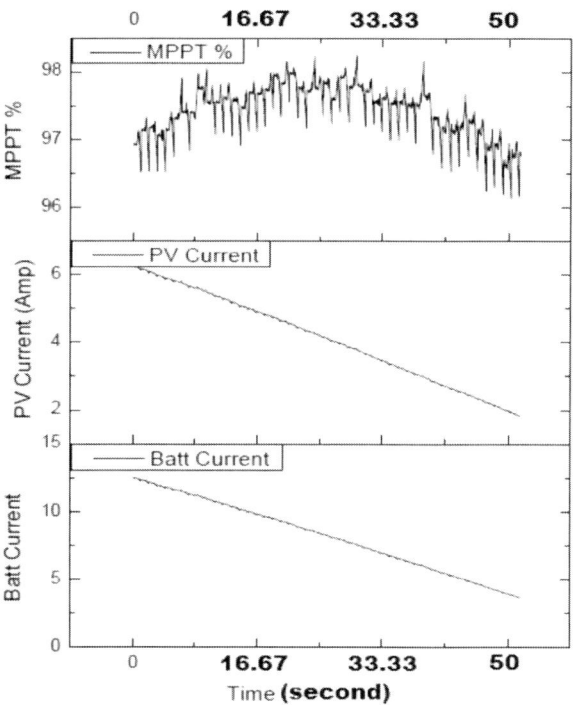

Figure 20. MPP tracking efficiency and PV and battery charging current in dynamic condition (when the irradiation changes from 100 W/m^2 to 500 W/m^2).

The second set, when the charge controller has been operated with a PV configuration of V_{oc} = 115V in the PV simulator. The last set, when the charge controller has been operated with a PV configuration of V_{oc} = 135V in the PV simulator. In each set of operations, the irradiation has been set in the PV simulator at that level which is required for generating output power according to the load percentage and the PV and battery parameters reading have been noted down to calculate the conversion efficiency. Besides the job of finding the power conversion efficiency, the MPP tracking efficiency has also been noted down during this whole experimental operation.

Table 4.2. Test results when tested @ PV voltage of 85V (Battery voltage less than 50V and temperature 25°C)

SL No.	Load%	5%	10%	25%	50%	75%	100%
1	Battery Voltage (V)	48.1	48.2	48.4	48.8	49.2	49.7
2	Battery Current (A)	1.33	2.92	7.49	14.88	21.98	29.88
3	Battery Power (W)	64.3	140.7	362.8	726.1	1081.33	1484.9
4	PV Voltage (V)	71.3	72.9	74.7	76.4	78.0	79.6
5	PV Current (A)	1.0	2.0	4.9	9.6	14.1	19.3
6	PV Power (W)	68.6	144.2	367.3	733.4	1097.8	1532.3
7	Converter efficiency (%)	93.8	97.6	98.8	99.0	98.5	97.5
8	MPP tracking efficiency (%)	95.7	96.8	98.8	99.2	98.6	97.5

Table 4.3. Test results when tested @ PV voltage of 115V (Battery voltage less than 50V and temperature 25°C)

SL No.	Load%	5%	10%	25%	50%	75%	100%
1	Battery Voltage (V)	48.2	48.3	48.5	49.0	49.3	49.7
2	Battery Current (A)	1.69	3.0	7.48	16.37	22.54	28.94
3	Battery Power (W)	81.8	145	362.85	802.2	1111.7	1438.4
4	PV Voltage (V)	96.4	95.5	100.1	112.1	101.6	103.1
5	PV Current (A)	0.9	1.6	3.7	7.3	11.2	14.2
6	PV Power (W)	86.9	149.8	369.5	816.9	1135.6	1469.3
7	Converter efficiency (%)	94.2	96.8	98.2	98.2	97.9	97.9
8	MPP tracking efficiency (%)	97.9	98.3	98.6	99.7	99.9	99.7

It has been observed in Figure 21(a), that the maximum converter efficiency of the charge controller has been found 99%. In Table 4.2, 4.3 and 4.4, the PV operating voltage for different loading conditions has been shown and it is found that the new charge controller is operating in a wide input voltage range i.e., 71.3V – 127.8V. And in Figure 21(b), the MPP tracking efficiency for different loading conditions has been shown and it is found that the highest MPP tracking efficiency has been found at 99.9%.

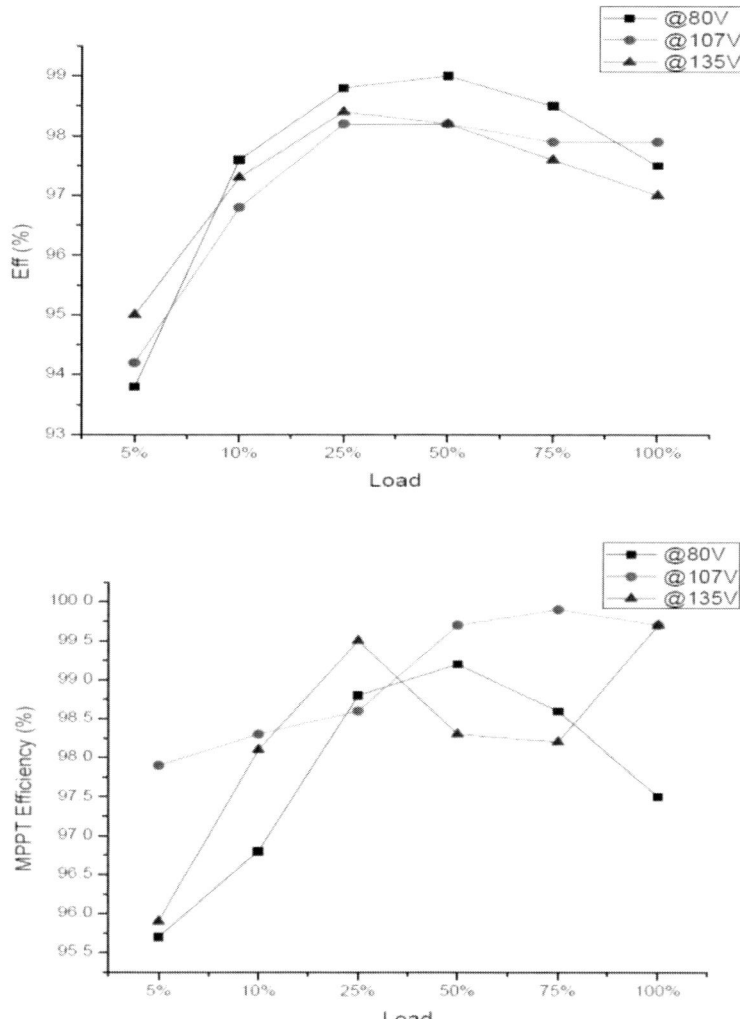

Figure 21. (a) Converter efficiency curve for different load conditions, (b) MPP tracking efficiency curve for different load conditions.

Table 4.4. Test results when tested @ PV voltage of 135V (Battery voltage less than 50V and temperature 25°C)

SL No.	Load%	5%	10%	25%	50%	75%	100%
1	Battery Voltage (V)	48.4	48.5	48.8	49.1	49.5	49.9
2	Battery Current (A)	1.5	3.14	7.91	15.49	23.26	30.03
3	Battery Power (W)	72.8	152.6	386.12	760.36	1151.19	1498.3
4	PV Voltage (V)	106.4	116.8	120.9	128.6	125.4	127.8
5	PV Current (A)	0.7	1.3	3.2	6.0	9.4	12.1
6	PV Power (W)	76.6	156.8	392.4	774.3	1179.5	1544.7
7	Converter efficiency (%)	95	97.3	98.4	98.2	97.6	97
8	MPP tracking efficiency (%)	95.9	98.1	99.5	98.3	98.2	99.7

5. PERFORMANCE COMPARISON

A comparison between the developed MPPT charge controller and the TriStar MPPT charge controller manufactured Morningstar by has been presented in Table 5.1.

Table 5.1. Performance comparison of the developed MPPT charge controller with the TriStar MPPT charge controller manufactured by Morningstar

Parameter	Morningstar TriStar TS-MPPT-30	The developed MPPT charge controller
Nominal system voltage	12, 24, or 48 VDC	48 VDC
Maximum battery current	30 Amp	30 Amp
Maximum PV open circuit voltage	150 VDC	150 VDC
Maximum self-consumption	2.7 W	2.7 W
Battery charging algorithm	4-stage	5-stage
Battery charging stages	Bulk, absorption, float, equalization	Soft, bulk, absorption, float, equalization

6. COST ESTIMATION OF THE DEVELOPED CHARGE CONTROLLER

To find out the prototype making cost of the newly developed charge controller hardware, the bill-of-material (BOM) of the primary components has been listed in Table6.1.

Table 6.1. Specifications, pricing of the primary materials used in the developed charge controller

Type	Specifications	Quantity	Manufacturer	Footprint	Price (Rs.)
Resistance	47k, 500 mW, ± 5%, 200 V	1	TE CONNECTIVITY	1206 [3216 Metric]	2.3
	VARISTOR, 180V, 115VAC, 150VDC	1	BOURNS	Disc 14mm	1.5
	120 K, 250 mW, ± 1%, MF	1	MULTICOMP	MELF DIN 0204 [3715 Metric]	2.3
	100 K, 250 mW, ± 1%, MF	6	MULTICOMP	MELF DIN 0204 [3715 Metric]	14.4
	10 K, 250 mW, ± 1%, MF	3	MULTICOMP	MELF DIN 0204 [3715 Metric]	7.2
	82 K, 250 mW, ± 1%, MF	1	MULTICOMP	MELF DIN 0204 [3715 Metric]	2.4
	22E, 2W, 5%, 150 V	2	LOCAL	0414/15	
	10 K, 125 mW, ± 1%, 150 V	5	VISHAY	0805 [2012 Metric]	3.9
	22K, 250mW, 5%, Metal Film	1	LOCAL	1206 [3216 Metric]	
	220 E, 125 mW, ± 1%, 150 V	5	VISHAY	0805 [2012 Metric]	3.6
	10 ohm, 500 mW, ± 1%, 200 V	2	TE CONNECTIVITY	1206 [3216 Metric]	8.5
	0R, 1%, 0.5W, 2010	2	MULTICOMP	2010 [5025 Metric]	2
Capacitor	0.1 µF, ± 10%, X7R, 100 V	13	AVX Corporation	0805 [2012 Metric]	26.7
	0.1 µF, ± 10%, X7R, 250 V	2	LOCAL	C075-052X106	
	470uF/250V	4	LOCAL	EB22,5D	

Table 6.1. (Continued)

Type	Specifications	Quantity	Manufacturer	Footprint	Price (Rs.)
Capacitor	22nF/150V	2	LOCAL	13.4 x 6 mm, grid 10.16 mm	
	2200uF/100V	2	LOCAL	EB25D	
	0.1 µF, ± 10%, X7R, 100 V	3	LOCAL	C050-025X075	
	1 µF, ± 10%, X7R, 16 V	2	KEMET	0805 [2012 Metric]	6
	10uF/25V	2	PANASONIC	PANASONIC_C	48
	2.2 µF, ± 10%, 25 V	1	VISHAY	1411 [3528 Metric]	41.24
Diode	1N4007	4	SYC	D0-214 AC	
	TVS, SMAJ Series, 150 V, 167 V	1	MULTICOMP	D0-214 AA	17.4
	200 V, 60 A, 1.35 V, 30 ns, 250 A Ultrafast Diode	4	STMICRO ELECTRONICS	TO-247	1,074
	TVS DIODE, 600W, 64V	1	FAIRCHILD SEMICONDUCTOR	D0-214 AA	32.82
	Small Signal Schottky Diode	3	NXP	SOD-123F	24
MOSFET	MOSFET Transistor, N Channel	4	Infineon Technologies	TO247	2,325.84
	BD139	1	LOCAL		
Gate Driver	HCPL-3120-000E	2	BROADCOM LIMITED	DIP 8	622.44
	IRS2186SPBF	1	Infineon Technologies	SOIC 8	198.84
Hall Sensor	ACS758LCB-050U-PFF-T	2	Allegro MicroSystems	CB 5-pin	1,481.06
DC-DC Converter	IL1212S	1	XP Power	Through Hole	611.66
Heatsink	Diode Heatsink	3	AAVID THERMALLOY	35 mm, 35 mm, 22 mm	603.6
	IGBT Heatsink	1	AAVID THERMALLOY	54mm *38mm *100mm	714.65
Fuse & Holder	30 A, Fast Acting	1	LITTELFUSE	19.1mm x 5mm x 19mm	19.2
	40 A, Fast Acting	1	LITTELFUSE	19.1mm x 5mm x 19mm	19.2
	Fuse Holder	2	MULTICOMP	5mm x 20mm	122
Connector	LAMA Connector	4	PANDUIT		241.48

Type	Specifications	Quantity	Manufacturer	Footprint	Price (Rs.)
Fastener	SCREW SOCKET		TR FASTENINGS		559
	NYLOC NUT, M6		DURATOOL		838
	PLAIN WASHER, M6		DURATOOL		357
	SPR WASHER, M6		DURATOOL		217
	S/PROOF WASHER, M6		DURATOOL		183
Total					10,432.23

CONCLUSION

The battery charge controller for operation in Battery Back-up Grid-Import PV PCU should be a combined solution for maximizing the transfer of solar electricity generated by the PV module(s) and also for ensuring a long battery life. But both these aspects of battery charge controller for Battery Back-up Grid-Import PV systems have generally been investigated independently in technical literature earlier and are presented in an integrated fashion in this article which highlights the scope of improvements in photovoltaic power conditioning units. Several MPPT techniques have been reviewed and presented here along with their salient features of operations. It briefly reviews the limitations of the existing charge controllers used in the Battery Back-up Grid-Import PCUs.

An improved charge controller has been designed and developed, which is a combined solution of a five-stage battery charging and an efficient MPPT technique. Two additional stages have been included in the charge controller namely, the soft charging and the equalization charging with the three existing stages of traditional charge controllers which ensures a balance between battery overcharging and undercharging and prevents battery stratification resulting in a longer battery life. The charge controller employs an improved P&O algorithm with a variable duty cycle for fast MPP tracking and very low PV operating point fluctuation around the MPP, and ensures a high static and dynamic MPP tracking efficiency. Also, the incompatibility issues like the large range of input PV voltage and temperature variation has been investigated and taken into consideration during the design process.

The experimental results show that the new charge controller has maximum converter efficiency of 99%, and it can extract as much power as possible to charge the battery with an MPP tracking speed of 1 second and maximum MPP tracking efficiency of 99.9% with fluctuation of 0.86% around the target MPP in static irradiation condition. And in dynamic irradiation conditions, the before mentioned performance parameters become 1.67 seconds, 98.5% and 0.50% respectively. The charge controller has a very wide input voltage range of 71.3V – 127.8V with which it can be coupled to a wide range of PV modules available in today's market for charging batteries in the PV application.

REFERENCES

[1] Chiandone, M., Tam, C., Campaner, R., Sulligoi. G. 2017, "Electrical storage in distribution grids with renewable energy sources. In: *International Conference on Renewable Energy Research and Applications* (ICRERA)," 2017.

[2] Desconzi, M., Beltrame, R., Rech, C., Schuch, L., Hey, H., "Photovoltaic standalone power generation system with multilevel inverter," *Renewable Energy & Power Quality Journal*, Vol. 1, No. 9, May 2011.

[3] Jana J, Bhattacharya KD, Saha H. 2014, "Design & implementation of MPPT algorithm for battery charging with photovoltaic panel using FPGA," 2014. *In: Power India International Conference (PIICON)*, 10.1109/POWERI.2014.7117704.

[4] Reisi Ali Reza, Morad Mohammad Hassan, Jamasb Shahriar, 2013, "Classification and comparison of maximum power point tracking techniques for photovoltaic system: a review." *Renew Sustain Energy Rev* 2013:475–88.

[5] Hirech, K., Melhaoui, M., Yaden, F., Baghaz, E., Kassmi, K., "Design and realization of an autonomous system equipped with a

charge/discharge regulator and digital MPPT command." *Energy Proc.* 42, 503–512, The Mediterranean Green Energy Forum 2013.

[6] Pamela G. Horkos, Emile Yammine, Nabil Karami, "*Review on Different Charging Techniques of Lead-Acid Batteries,*" 10.1109/TAEECE.2015.7113595.

[7] MdShahrier Hakim, Farhana Latif, Md. Imran Khan, Al Basir, "Design and implementation of three-stage battery charger for lead-acid battery," 10.1109/CEEICT.2016.7873052, *3rd International Conference on Electrical Engineering and Information Communication Technology* (ICEEICT), 2016.

[8] Salas V., E. Olı́as, A. Barrado, A. Laźaro, "Review of the maximum power point tracking algorithms for stand-alone photovoltaic systems," *Solar Energy Materials & Solar Cells* Volume 90, Issue 11, Page 1555–1578, July 2006.

[9] Esram T., Kimball J. W., Krein P. T., Chapman P. L., Midya P., "Dynamic maximum power point tracking of photovoltaic arrays using ripple correlation control," *IEEE Trans. Power Electron.*, vol. 21, no. 5, pp. 1282–1291, Sep 2006.

[10] Kazuhiro Kajiwara, Nobumasa Matsui, Fujio Kurokawa, 2017 "A new MPPT control for solar panel under bus voltage fluctuation. In: *International Conference on Renewable Energy Research and Applications* (ICRERA)," 2017.

[11] Rizzo, S. A., N. Salerno, G. Scelba, A. Sciacca, 2018 "Enhanced Hybrid Global MPPT Algorithm for PV Systems Operating under Fast-Changing Partial Shading Conditions," *International Journal of Renewable Energy Research*, 2018 Vol. 8, No. 1.

[12] Xingshuo Li, Huiqing Wen, Yihua Hu, 2016 "Evaluation of different maximum power point tracking (MPPT) techniques based on practical meteorological data. In: *International Conference on Renewable Energy Research and Applications* (ICRERA)," 2016.

[13] Priyadarshi, Neeraj, Amarjeet Kumar Sharma, S Priyam, 2017 "Practical Realization of an Improved Photovoltaic Grid Integration with MPPT," *International Journal of Renewable Energy Research*, 2017, Vol. 7, No. 4.

[14] Dolara, A., S. Leva, G. Magistrati, M. Mussetta, E. Ogliari, R. Varun Arvind, 2016 "A novel MPPT algorithm for photovoltie systems under dynamic partial shading — Recurrent scan and track method. In: *International Conference on Renewable Energy Research and Applications* (ICRERA)," 2016.

[15] Priyadarshi, Neeraj, Amarjeet Kumar Sharma, Faarooque Azam, 2017 "A Hybrid Firefly-Asymmetrical Fuzzy Logic Controller based MPPT for PV-Wind-Fuel Grid Integration," *International Journal of Renewable Energy Research*, 2017, Vol. 7, No. 4.

[16] Graditi, G., G. Adinolfi, A. Del Giudice, 2015 "Experimental performances of a DMPPT multitopology converter. In: *International Conference on Renewable Energy Research and Applications* (ICRERA)," 2015.

[17] López, Julio, S. I. Seleme Jr., P. F. Donoso, L. M. F. Morais, P. C. Cortizo, M.A. Severo, "Digital control strategy for a buck converter operating as a battery charger for stand-alone photovoltaic systems." *Sol. Energy* 140, 171–187, Dec 2016.

[18] Reddy T. Linden, 2010. Reddy T. *Linden's handbook of batteries.* 4th ed. McGraw-Hill Education; 2010.

[19] Parker D. A. J., Moseley P. T., Garche J., Parker C. D., 2004 *Valve-regulated lead-acid batteries.* Amsterdam: Elsevier; 2004.

[20] Hunter P. M., 2003 *VRLA battery float charge: analysis and optimization.* Christchurch, New Zealand: University of Canterbury; 2003.

[21] Wong Y., Hurley W., Wölfle W., 2008 "Charge regimes for valve-regulated lead-acid batteries: performance overview inclusive of temperature compensation." *J Power Sources* 2008; 183(2):783–91.

[22] Valeriote E. M., Nor J., Ettel V. A. "Very fast charging of lead-acid batteries." In: *Proceedings of the 5th international lead-acid battery seminar, International Lead Zinc Research Organization (ILZRO).* p. 93–122.

[23] Fleming F., Shumard P., Dickinson B., 1999 "Rapid recharge capability of valve regulated lead-acid batteries for electric vehicle

and hybrid electric vehicle applications." *J Power Sources* 1999;78(1):237–43.
[24] Hunter P. M., Anbuky A. H. 2003 "VRLA battery rapid charging under stress management." *IEEE Trans Industr Electron* 2003;50(6):1229–37.
[25] Jones R. H., McAndrews J. M., Vaccaro F., 1998 "Recharging VRLA batteries for maximum life. In: Twentieth international telecommunications energy conference," 1998. *INTELEC. IEEE; 1998*. p. 526–31.
[26] Li Y., Chattopadhyay P., Ray A., 2015 "Dynamic data-driven identification of battery state-of-charge via symbolic analysis of input–output pairs." *Appl Energy* 2015;155:778–90.
[27] Boisvert É., 2001 "Using float charging current measurements to prevent thermal runaway on VRLA batteries." In: *Twenty-third international telecommunications energy conference, 2001. INTELEC 2001. IET*; 2001. p. 126–31.
[28] Klein R., Chaturvedi N. A., Christensen J., Ahmed J., Findeisen R., Kojic A., 2011 "Optimal charging strategies in lithium-ion battery. In: American Control Conference (ACC)," 2011. *IEEE; 2011*. p. 382–7.
[29] Fathabadi H., 2015 "Lambert W function-based technique for tracking the maximum power point of PV modules connected in various configurations." *Renewable Energy* 2015;74:214–26.
[30] Tsang K. M., Chan W. L., 2013 "Model based rapid maximum power point tracking for photovoltaic systems." *Energy Convers Manage* 2013;70:83–9.
[31] Tsang K. M., Chan W. L., "Three-level grid-connected photovoltaic inverter with maximum power point tracking." *Energy Convers Manage* 2013;65:221–7.
[32] Ammar M. B., Chaabene M., Chtourou Z., 2013 "Artificial Neural Network based control for PV/T panel to track optimum thermal and electrical power." *Energy Convers Manage* 2013;65:372–80.
[33] Abdelsalam A. K., Massoud A. M., Ahmed S., Enjeti P. N., 2011 "High-performance adaptive perturb and observe MPPT technique

for photovoltaic-based micro grids." *IEEE Trans Power Electron* 2011;26(4):1010–21.

[34] Brunton S. L., Rowley C. W., Kulkarni S. R., Clarkson C., 2010 "Maximum power point tracking for photovoltaic optimization using ripple-based extremum seeking control." *IEEE Trans Power Electron* 2010;25(10):2531–40.

[35] Dileep G., S. N. Singh, "Application of soft computing techniques for maximum power point tracking of SPV system," *Sol. Energy* 141, 182–202, Jan 2017.

[36] Enslin J. H. R., Wolf M. S., Snyman D. B., Swiegers W., 1997 "Integrated photovoltaic maximum power point tracking converter." *IEEE Trans Ind Electron* 1997;44:769–73.

[37] Hiyama T., Kitabayashi K., 1997 "Neural network based estimation of maximum power generation from PV module using environmental information." *IEEE Trans Energy Convers* 1997;12:241–7.

[38] Jiang J. A., Huang T. L., Hsiao Y. T., Chen C H., 2005 "Maximum power tracking for photovoltaic power systems." *J Sci Eng* 2005;8(2):147–53.

[39] Lei P., Li Y., Seem J. E., 2011 "Sequential ESC-based global MPPT control for photovoltaic array with variable shading," *IEEE Trans Sustain Energy* 2011;2(3):348–58.

[40] Liu F., Duan S., Liu F., Liu B., Kang Y., 2008. A variable step size INC MPPT method for PV systems. *IEEE Trans Ind Electron* 2008;55(7):2622–8.

[41] Masoum M. A. S., Dehbonei H., Fuchs E. F., 2002 "Theoretical and experimental analyses of photovoltaic systems with voltage and current-based maximum power-point tracking." *IEEE Trans Energy Convers* 2002;17(4):514–22.

[42] Mei Q., Shan M., Liu L., Guerrero J. M., 2011 "A novel improved variable step-size incremental-resistance MPPT method for PV systems." *IEEE Trans Ind Electron* 2011;58(6):2427–34.

[43] Noguchi T., Togashi S., Nakamoto R., 2002. "Short-current pulse-based maximum power-point tracking method for multiple

photovoltaic-and-converter module system." *IEEE Trans Ind Electron* 2002;49(1):217–23.

[44] Park M., In-Keun Yu, 2004 "A study on the optimal voltage for MPPT obtained by surface temperature of solar cell." *In: 30th annual conference of IEEE* 3; 2004. p. 2040–5.

[45] Al Nabulsi Ahmad, Rached Dhaouadi., 2012 "Efficiency optimization of a DSP-based standalone PV system using fuzzy logic and dual-MPPT control," *IEEE Trans Ind Inform* 2012;8(3):573–84.

[46] Hsieh Guan-Chyun, Hung-IHsieh, Tsai Cheng-Yuan, Wang Chi-Hao, 2013 "Photovoltaic power-increment-aided incremental-conductance MPPT with two-phased tracking." *IEEE Trans Power Electron* 2013;28(6):2895–911.

[47] Abdul Rahman N. H., Omar A. M., Mat Saat E. H., 2013 "A modification of variable step size INC MPPT in PV system. In: *Proceedings of the IEEE 7th international conference on power engineering and optimization*"; 2013.p.340–345.

[48] Algazar M. M., AL-monier H., EL-halim H. A., El Kotb Salem ME, 2012 "Maximum power point tracking using fuzzy logic control." *Int J Electr Power Energy Syst* 2012;39 (1):21–8.

[49] Guenounou O., Dahhou B., Chabour F., 2014. Adaptive fuzzy controller based MPPT for photovoltaic systems. Energy Convers Manage 2014;78:843–50.

[50] Mellit A., Sag˜lam S., Kalogirou S. A., 2013 "Artificial neural network-based model for estimating the produced power of a photovoltaic module." *Renewable Energy* 2013;60:71–8.

[51] Bazzi A. M., Krein P. T., 2011 "Concerning maximum power point tracking for photovoltaic optimization using ripple-based extremum seeking control." *IEEE Trans Power Electron* 2011;26(6):1611–2.

[52] Alabedin, A. M. Z., El-Saadany, E. F., Salama, M. M. A., "Maximum power point tracking for photovoltaic systems using fuzzy logic and artificial neural networks." *In: Power and Energy Society General Meeting. IEEE*, pp. 1–9, 2011.

[53] Jinbang, X., Anwen, S., Cheng, Y., Wenpei, R., Xuan, Y., 2011 "ANN based on IncCond algorithm for MPP tracker". *In: Bio-*

Inspired Computing: Theories and Applications (BIC-TA). 2011 Sixth International Conference, pp. 129–134.

[54] Ramaprabha, R., Mathur, B. L., "Intelligent controller based maximum power point tracking for solar PV system. *Int. J. Comput. Appl.* (0975–8887) 12 (10), 2011.

[55] Messai, A., Mellit, A., Guessoum, A., Kalogirou, S. A., "Maximum power point tracking using GA optimize fuzzy logic controller and its FPGA implementation." *Sol. Energy* 86, 265–277, Feb 2011.

[56] Subiyanto, Mohamed, A., Shareef, H., "Neural network optimized fuzzy logic controller for maximum power point tracking in a photovoltaic system." *Int. J. Photoenergy*, 6–13, 2012.

[57] Khaehintung, N., Pramotung, K., Tuvirat, B., Sirisuk, P., "RISC-microcontroller built-in fuzzy logic controller of maximum power point tracking for solar powered light-flasher applications." *Indust Electron Soc, IECON 30th Annu Conf IEEE*, vol. 3, pp. 2673–2678, 2004.

[58] Alajmi, B. N., Ahmed, K. H., Finney, S. J., Williams, B. W., 2011. "Fuzzy-logic-control approach of a modified hill-climbing method for maximum power point in microgrid standalone photovoltaic system." *IEEE Trans. Power Electron.* 26, 1022–1030, 2010.

[59] Kchaou A., Naamane A., Koubaa Y., M'sirdi N., "Second order sliding mode-based MPPT control for photovoltaic applications," *Sol. Energy* 155, 758–769, 2017.

[60] Li Shaowu, Liao Honghua, Yuan Hailing, Ai Qing, Chen Kunyi, "A MPPT strategy with variable weather parameters through analyzing the effect of the DC/DC converter to the MPP of PV system, *Sol. Energy* 144, 175–184, 2017.

[61] Mohamed Orabi, Ahmed Shawky, 2015 "Proposed Switching Losses Model for Integrated Point of Load Synchronous Buck Converters," *IEEE Transactions on Power Electronics* 2015; 30(9): 5136 – 5150.

[62] Al-Diab, A., C. Sourkounis, 2010 "Variable step size PO MPPT algorithm for PV systems," in: *Proceedings of Optimization of Electrical and Electronic Equipment (OPTIM) Conference*, 2010, pp. 1097–1102.

[63] Safari A, Mekhilef S, 2011 "Simulation and hardware implementation of incremental conductance MPPT with direct control method using cuk converter." *IEEE Trans. Ind. Electron.* 2011;58:1154–61.

[64] Al-Diab, A., C. Sourkounis, 2010 "Variable step size PO MPPT algorithm for PV systems," in: *Proceedings of Optimization of Electrical and Electronic Equipment (OPTIM) Conference*, 2010, pp. 1097–1102.

[65] Fangrui L., Shanxu D., Fei L., Bangyin L., Yong K., 2008 "A variable step size INC MPPT method for PV systems." *IEEE Trans. Ind. Electron.* 2008;55:2622–8.

[66] Harrag, Abdelghani. Sabir Messalti, 2015 "Variable step size modified P&OMPPT algorithm using GA-based hybrid offline/online PID controller," *Renewable and Sustainable Energy Reviews* 49(2015)1247–1260.

[67] Khadidja Saidi, Mountassar Maamoun, M'hamed Bounekhla, 2017 "Simulation and analysis of variable step size P&O MPPT algorithm for photovoltaic power control," *2017 International Conference on Green Energy Conversion Systems (GECS)*.

[68] Peng, B. R., J. H. Chen, Y. H. Liu, Y. H. Chiu, 2015 "Comparison between Three Different Types of Variable Step-Size P&O MPPT Technique," *International Conference on Computer Information Systems and Industrial Applications (CISIA 2015)*.

[69] Joydip Jana, Hiranmay Samanta, Konika Das Bhattacharya, Hiranmay Saha, "Design and development of high efficiency five stage battery charge controller with improved MPPT performance for Solar PV Systems," *International Journal of Renewable Energy Research*, Vol 8, No 2, 2018.

[70] EN 50530. *Overall efficiency of grid connected photovoltaic inverters.*

In: Maximum Power Point Tracking
Editor: Maurice Hébert
ISBN: 978-1-53618-164-7
© 2020 Nova Science Publishers, Inc.

Chapter 3

COMPARISON BETWEEN SiC- AND Si-BASED INVERTERS EQUIPPED WITH MAXIMUM POWER POINT TRACKING CHARGE CONTROLLER FOR PHOTOVOLTAIC POWER GENERATION SYSTEMS

Takeo Oku[1,*], Yuji Ando[1], Taisuke Matsumoto[1] and Masashi Yasuda[2]

[1]Department of Materials Science, The University of Shiga Prefecture, Hikone, Shiga, Japan
[2]Collaborative Research Center, The University of Shiga Prefecture, Hassaka, Hikone, Shiga, Japan

---

[*] Corresponding Author's E-mail: oku@mat.usp.ac.jp.

ABSTRACT

Power storage systems in ~100 W level were developed, which cosisted of direct current-alternating current converters, spherical Si solar cells, a maximum power point tracking controller, and lithium-ion batteries. Two types of inverters were used for the study: one is SiC metal-oxide-semiconductor field-effect transistors (MOSFETs) as switching devices while the other is conventional Si MOSFETs. In the present 100 W level inverters, the *on*-resistance was considered to have little influence on the efficiency. However, the SiC-based inverter exhibited an approximately 3% higher direct current-alternating current conversion efficiency than the Si-based inverter. Power loss analysis indicated that the higher efficiency resulted predominantly from lower switching and reverse recovery losses in the SiC MOSFETs compared with those in the Si MOSFETs.

Keywords: maximum power point tracking, silicon carbide, inverter, converter, solar cell, Photovoltaic device, Lithium-ion battery

1. INTRODUCTION

Requirements of recent power devices include a high blocking voltage, a low *on*-resistance, a high switching frequency, and good reliability. These requirements have led to great interest in power devices based on wide gap semiconductors such as GaN and SiC [1-9]. The main advantage of wide gap semiconductors is their very high electric field capability. Wide gap semiconductors possess high critical field strengths, which means that a thinner epitaxial layer is required to block the same voltage compared with Si devices. Thus, switching devices with much lower *on*-resistances can be fabricated using GaN or SiC. A lower *on*-resistance improves the efficiency of inverters due to reduced conduction and switching losses, and also decreases the module size due to the increased power density. The high electron mobility of GaN allows switching operations with higher frequencies, which also decreases the module size because of the smaller passive components. The excellent thermal stability of SiC and GaN should enable the devices to operate at high temperatures.

The improvement in system performance achieved by an improvement in device performance is an important consideration for such devices. Considerable researches have focused on developing power converters using SiC devices in an attempt to solve the problems [10-18]. A comparative study for Si- and SiC-based power converters was also reported [19-21]. In these papers, the power converters were examined as a stand-alone equipment. Concerning photovoltaic power applications, however, performances of total systems using these converters should be also compared.

In the present article, Si- and SiC-based inverters were compared with regard to performance of the total system including a maximum power point tracking (MPPT) controller, solar cell panels, and a storage battery as well as the inverter. We developed photovoltaic power generation systems based on two different direct current (DC)-alternating current (AC) converters. One inverter contained SiC metal-oxide-semiconductor field-effect transistors (MOSFETs) as switching devices while the other inverter contained Si MOSFETs. Stabilities of electrical power and DC-AC conversion efficiencies of the SiC- and Si-based systems were measured and compared. The compositions of power losses were also analyzed to understand the physical origin of the conversion losses in these systems. The capacity of handling power was set at around 100 W to enable the portability of the system. Lightweight spherical Si cell panels [22-29] were used as the power source. Spherical Si cells that are lightweight, flexible, and economical are suitable for self-sustaining energy systems such as portable electronic devices, solar-powered cars, and emergency power supply systems [30-38]. The present study also aimed to evaluate the feasibility of the photovoltaic power generation system based on spherical Si cells and the SiC-based inverter [39].

2. EXPERIMENTAL SETUP

A schematic model of the photovoltaic power generation systems based on two different DC-AC converters is shown in Figure 1 [19-21, 39,

40]. One inverter contained Si MOSFETs as switching devices (Daiji Industry, SXCD-300) while the other contained SiC MOSFETs. The SiC-based inverter was prepared by replacing Si MOSFETs (Fairchild, FQPF16N25C) [41] in SXCD-300 with SiC MOSFETs (Rohm, SCT2120AF) [42].

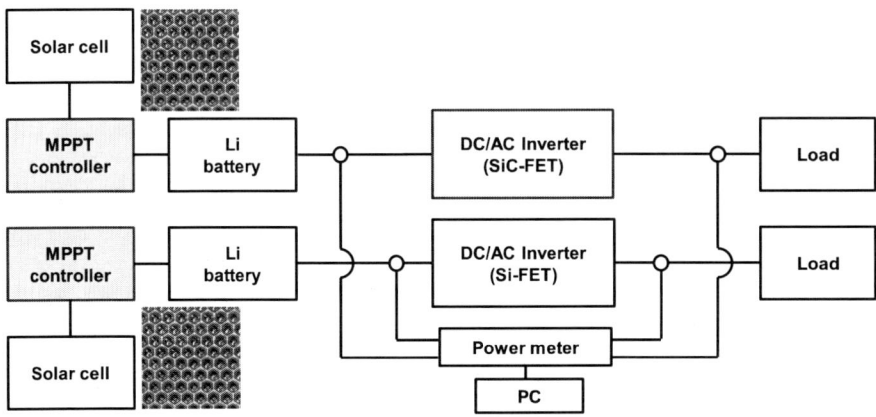

Figure 1. Schematic model of the photovoltaic power generation systems consisting of MPPT, spherical Si solar cells, and SiC-FET inverter.

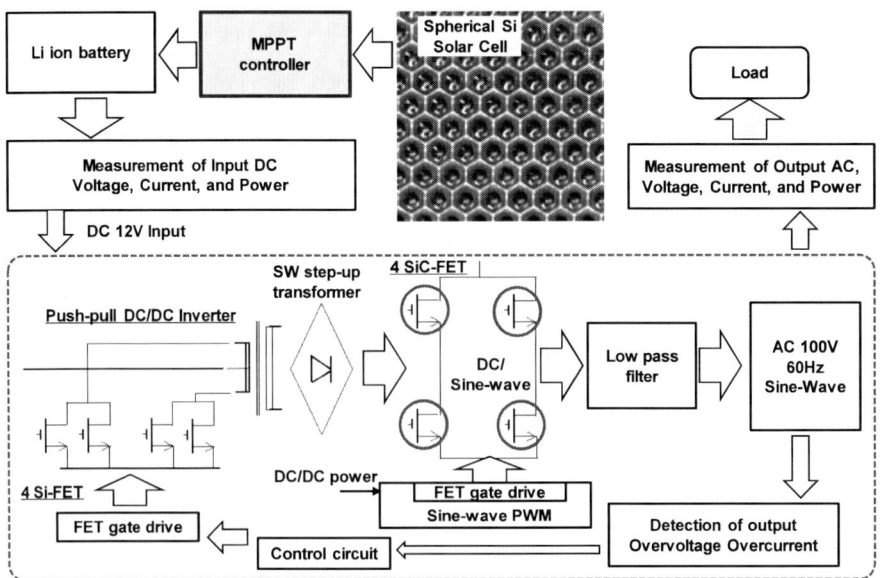

Figure 2. Schematic illustration of the photo-generation system with SiC-FET inverter.

Table 1. Specifications of the present inverters

	Si inverter	SiC inverter
Device	Si-FET FQPF16N25C	SiC-FET SCT2120AF
Size	147×66×210 mm	336×72×220 mm
Input	DC 12 V	DC 12 V
Output	AC 100 V	AC 100 V
Rated power	300 W	300 W
On-resistance max	270 mΩ	156 mΩ
On-resistance typ	220 mΩ	120 mΩ

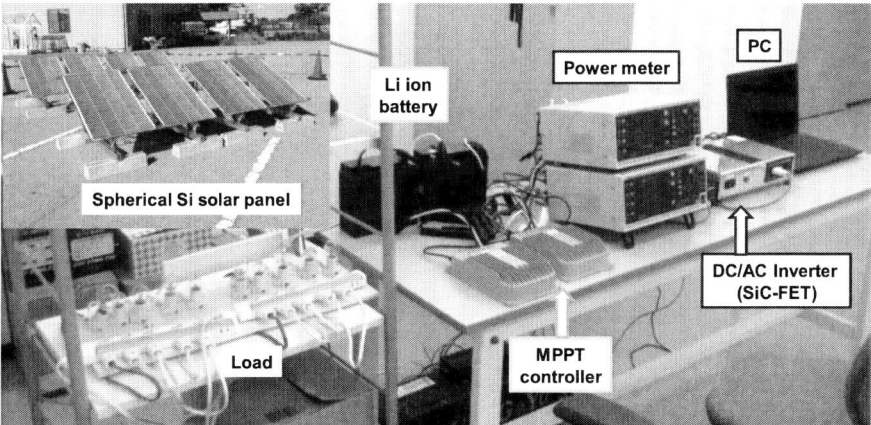

Figure 3. Digital photograph of experimental set up of the present photovoltaic power generation systems consisting of spherical Si solar cells, MPPT and SiC-FET inverter.

The inverter consisted of a front stage DC-DC converter followed by a second stage DC-AC converter, as shown in Figure 2. The Si MOSFETs of the DC-DC part exhibited a very low *on*-resistance of 4 mΩ, so SiC MOSFETs were introduced only in the DC-AC part, as listed in Table 1.

The input and output powers of these inverters were monitored synchronously by two power meters (Hioki, PW3336), as shown in Figure 3. The measurement interval was 200 ms and reported data are the average measurements over each minute. The spherical Si solar cell panels (Clean venture 21, CVFM-0540T2-WH) wired in parallel were used as the power source. The maximum operating current and voltage for a single solar panel were 3.34 A and 16.2 V, respectively. Detailed specifications of the spherical Si solar cells and Li ion battery are listed in Table 2 [39, 40]. The operating point was controlled to maximize the output power, with the aid of an MPPT controller (EPsolar, Tracer-2215BN). The MPPT stabilized the electricity by charging and discharging Li-ion batteries (O'Cell, IFM12-200E2). Filament lamps were used as the load. The temperature, humidity (Hioki, LR5001) and solar radiation power (Uizin, UIZ-PCM01-LR) were monitored simultaneously during measurements, as shown in Figures 1 and 3.

Table 2. Specifications of the present solar cells and Li ion battery

Solar cell	
Size	1020×690×5 mm
Weight	3.0 kg
Maximum power	54 W
Maximum output voltage	16.2 V
Maximum output current	1.67 A
Li ion battery	
Size	181×75×166 mm
Weight	2.75 kg
Rated capacity	DC 12 V, 20 Ah
Type	LiFePO$_4$

3. MAXIMUM POWER POINT TRACKING

There is a maximum energy output point on the current-voltage (*I-V*) curve owing to the nonlinear characteristics of the solar array. Conventional controllers with switch charging technology and pulse-width modulation (PWM) work by reducing the amount of power applied to the batteries as they get closer to be fully charged. Although the cost of the PWM controller is lower than the MPPT one, the efficiency is also lower, and the ranges are from 65~85%. The PWM system cannot charge the battery at the maximum power point to harvest the maximum energy available from PV array. On the other hand, the MPPT system is able to keep the point to harvest the maximum energy and deliver it to the battery. The MPPT algorithm adjusts the operating points to sustain the maximum power point of the array fully automatically, as shown in the *I-V* curve of Figure 4.

The present MPPT controller has a 3 efficient stages battery charging algorithm [43]. When the battery voltage has not yet reached constant voltage, the controller operates in constant current mode, delivering its maximum current to the batteries, which is called bulk charging or MPPT charging stage. When the battery voltage reaches the constant voltage setpoint, the MPPT controller will start to operate in a constant charging stage. The charging current will drop gradually to avoid too much gas precipitation or overheating of battery. After the constant voltage stage, the MPPT controller will reduce charging current to float voltage setpoint. This floating stage with a smaller voltage and current will have no more chemical reactions and all the charge current transforms into heat and gas at this time, while maintaining full battery storage capacity.

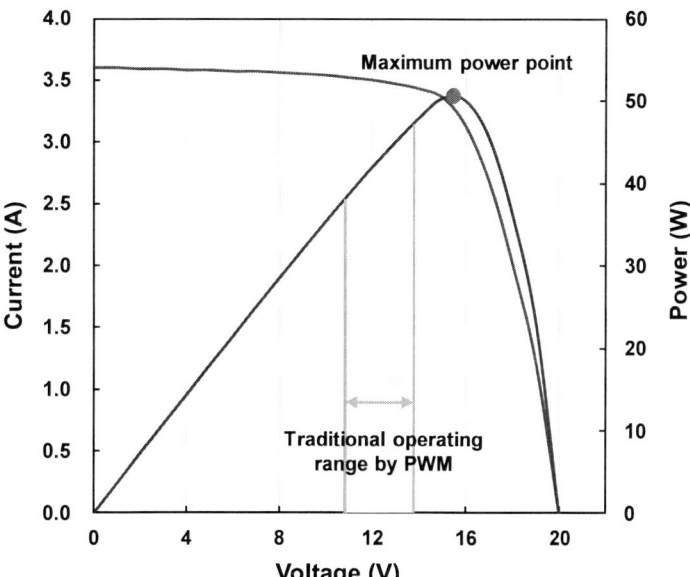

Figure 4. *I-V* characteristic and maximum power point.

Table 3. Specifications of the present MPPT controller

Property	Value
Size	$217 \times 143 \times 56$ mm^3
Weight	1.5 kg
Rated current	20 A
Maximum input	260 W (12 V)
Self-consumption	≤ 60mA (12V)
Efficiency of advanced MPPT	> 99.5%
Maximum conversion efficiency	~98%
Ambient temperature	-35 °C ~ +55 °C
Storage temperature	-35 °C ~ +80 °C
Humidity range	≤ 95%
Discharge circuit voltage drop	≤ 0.15V
Temperature compensate coefficient	-3 mV / °C / 2 V

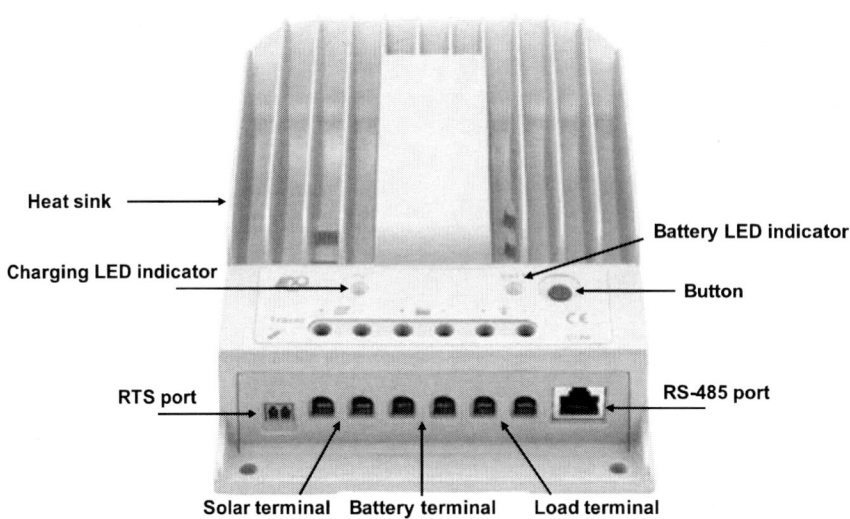

Figure 5. Digital photograph and functions of the present MPPT device.

A photographic image of the present MPPT device is shown in Figure 5. The device has die-cast aluminum design, ensuring excellent heat dissipation characteristic and light-emitting diode indicators showing system status. Real-time data checking and parameter settings are available for PC monitoring and external display unit connecting with RS-485 communication bus interface.

4. STABILITY OF ELECTRICITY

Figures 6(a), 6(b), 7(a), and 7(b) show examples of the variation in voltage, current, power, and DC-AC conversion efficiency, respectively, for the SiC- and Si-based inverters. The load power was set at 90 W. Figures 8(a) and 8(b) show the variation in solar radiation power and in temperature and humidity, respectively, during the measurements shown in Figure 7. In accordance with lowering of the solar radiation power after 14:30, a voltage reduction of the Li-ion battery became significant in the Si inverter system, which resulted in a gradual decrease in the input voltage (V_{in}). However, the input power (P_{in}) was almost kept constant due to the constant load power, and hence, the input current (I_{in}) was gradually increased. This kind of behavior was less significant in the SiC inverter system where the conversion efficiency was higher than that of the Si inverter system. Concerning the output voltage (V_{out}), current (I_{out}), and power (P_{out}), no fluctuation was observed in both systems. The DC-AC conversion efficiency increased from 83.4 to 86.5% using the SiC-based inverter. Thus, losses were reduced from 16.6 to 13.5% using the SiC-based inverter. This efficiency improvement in the SiC inverter was considered as a result of reduced switching and reverse recovery losses in the SiC MOSFETs, which are discussed in the next section.

Figures 9(a) and 9(b) show another example of the variation in power and DC-AC conversion efficiency, respectively, for Si- and SiC-based inverters. The load power was set at 10 W. In this case, discharge of the Li-ion battery was hardly observed. The currents (I_{in} and I_{out}) and voltages (V_{in} and V_{out}) were almost constant throughout a whole measurement. However, an initial decrease was observed in the input power in Figure 9 (a), which would be due to an initial variation of the efficiency. Besides, there was no significant difference in the efficiencies of the SiC- and Si-based inverters, as shown in Figure 9(b). Under low output power conditions ($P_{out} < 20$ W), the efficiency was predominantly determined by the non-load losses due to the transformers in the DC-DC converter circuit and the electrolytic capacitors of the filter circuit. Since the present SiC inverter was provided by just replacing a Si switch in the Si inverter with a SiC switch, non-load losses were considered to be invariable between them, and hence, no signifi-cant difference was expected in the efficiencies under low output power conditions.

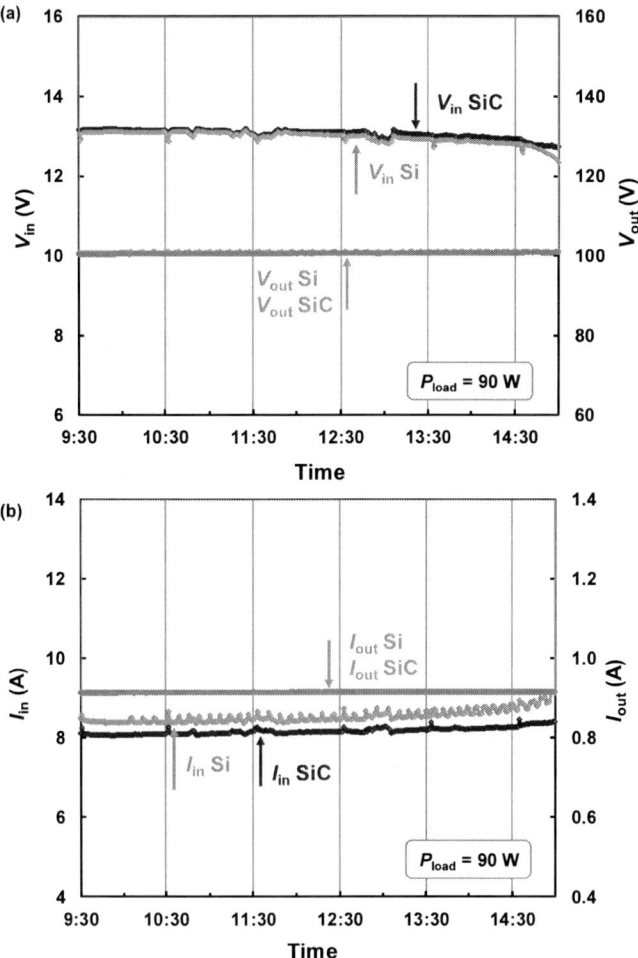

Figure 6. Variation in (a) voltage and (b) current at input and output terminals of Si and SiC-based inverters at a load power of 90 W.

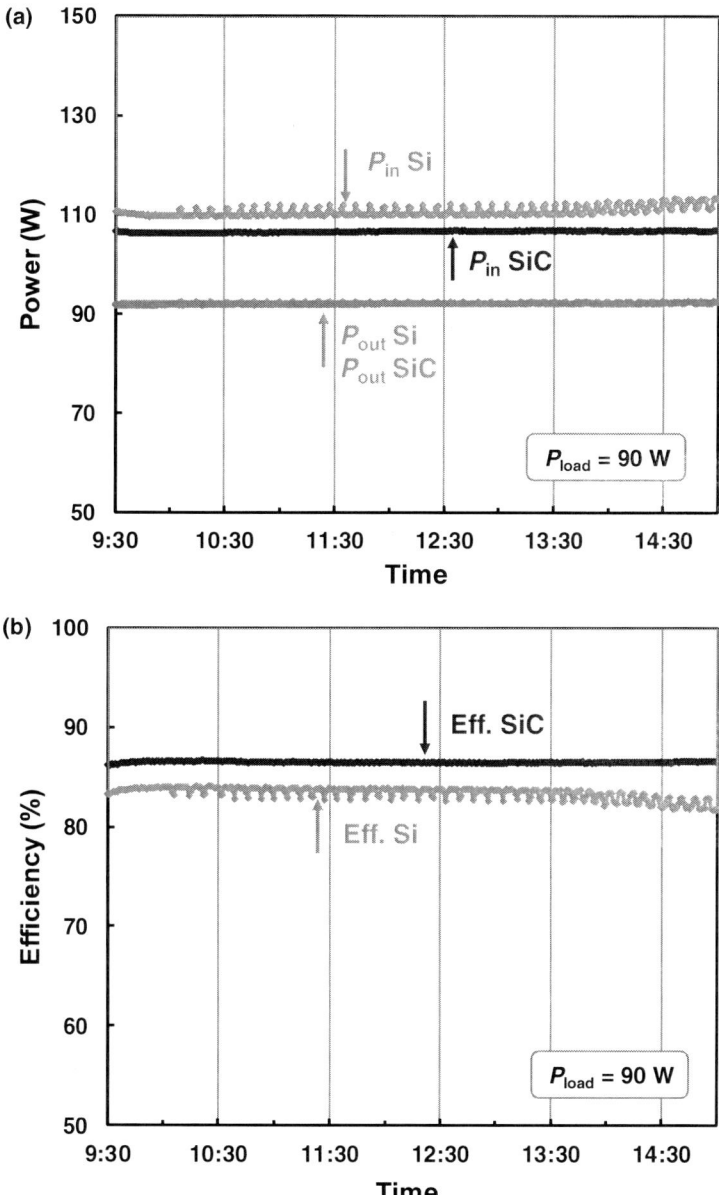

Figure 7. Variation in (c) power at input and output terminals and (d) DC-AC conversion efficiencies of Si and SiC-based inverters at a load power of 90 W.

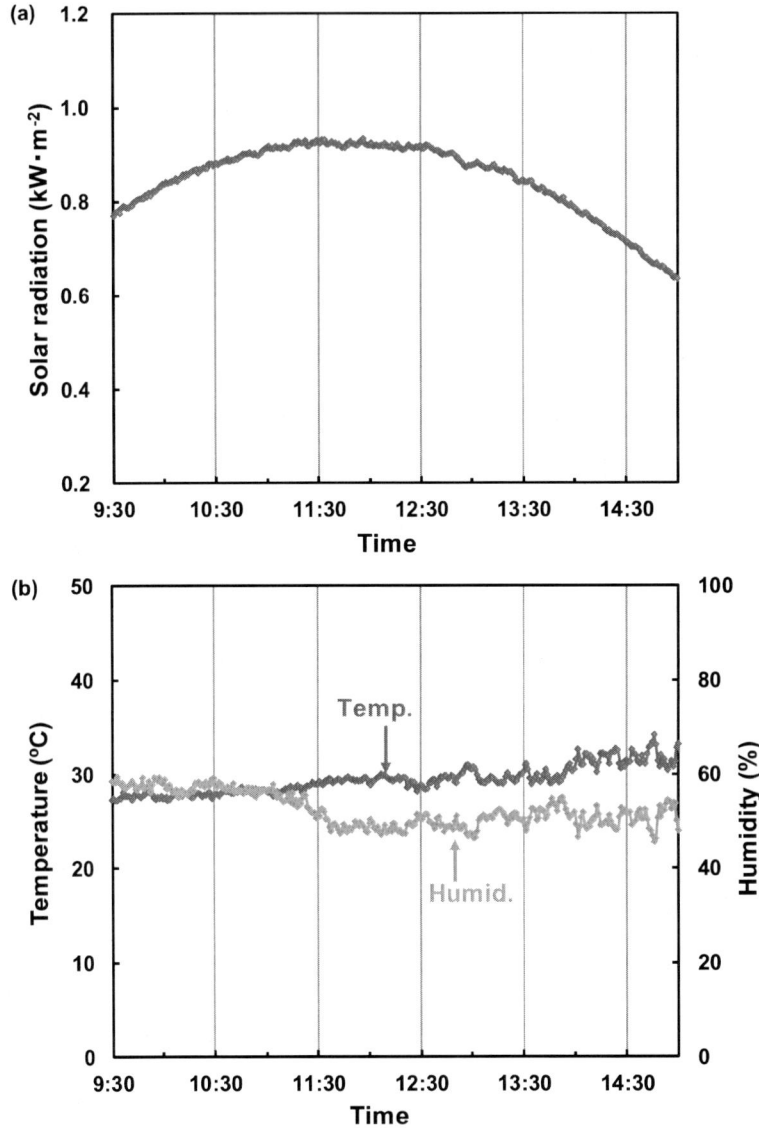

Figure 8. Variation in (a) solar radiation power and (b) temperature and humidity, during the measurements shown in Figure 7.

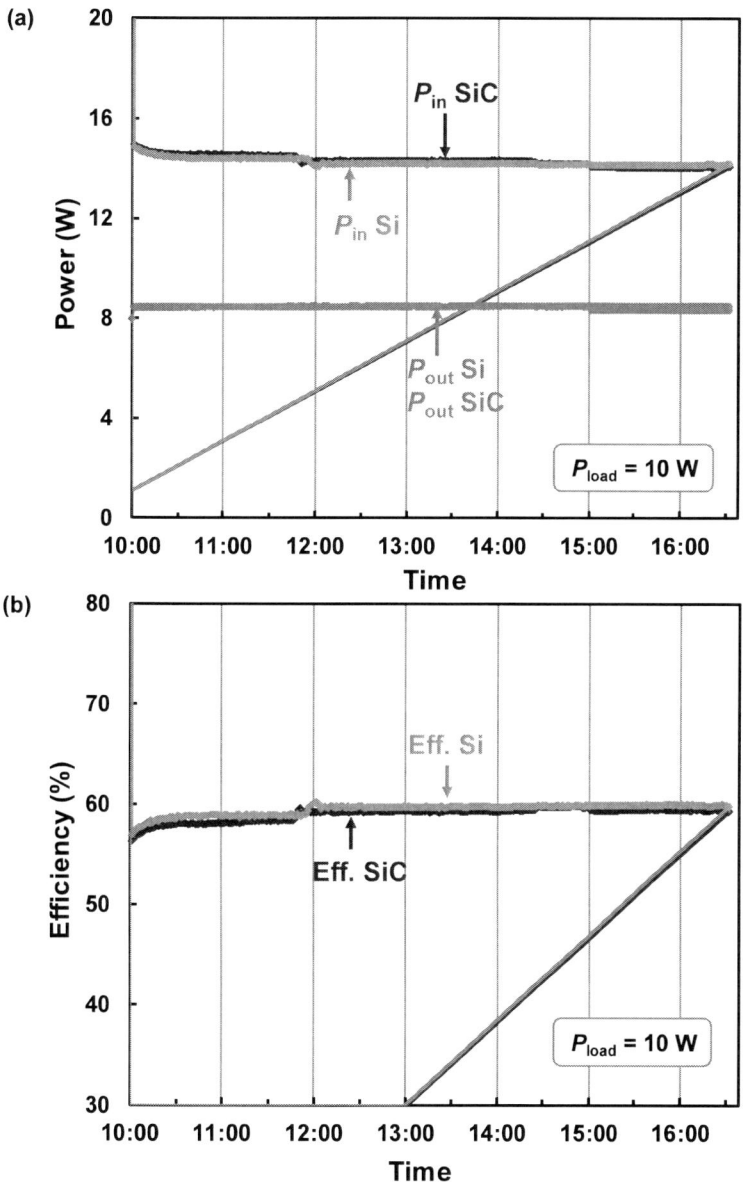

Figure 9. Variation in (a) input and output powers and (b) DC-AC conversion efficiencies of Si- and SiC-based inverters, at a load power of 10 W.

5. CONVERSION EFFICIENCIES

To study the influence of output power on the DC-AC conversion efficiencies, two different measurements were carried out.

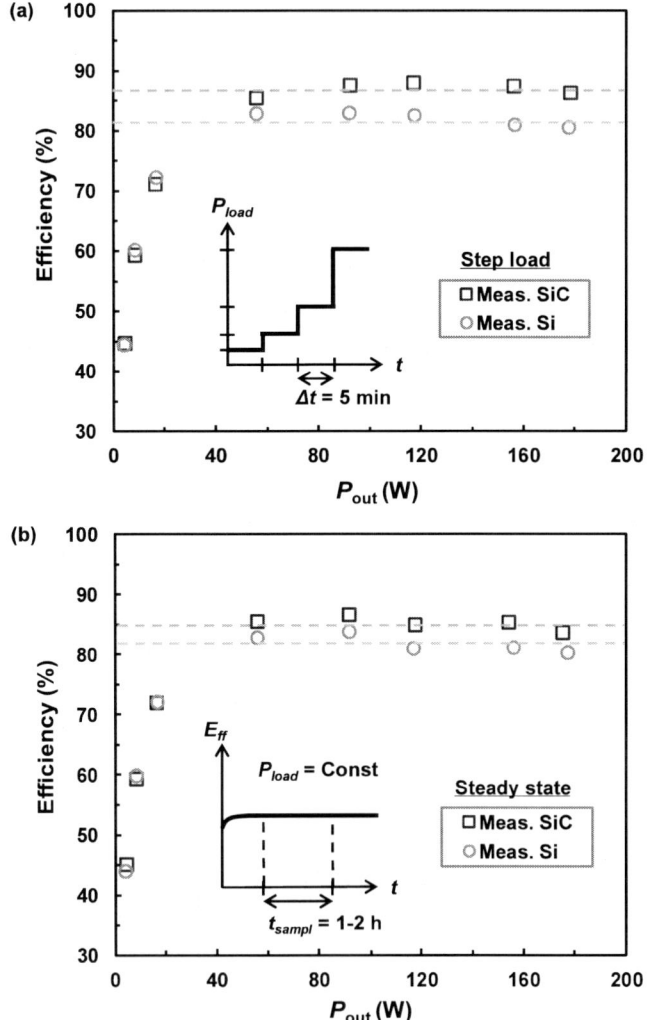

Figure 10. DC-AC conversion efficiencies versus output power for Si- and SiC-based inverters measured under (a) step load and (b) steady state conditions.

One was the step load measurement, where the load power was increased stepwise from 5 to 179 W, and held at each step for 5 min. The other was the steady state measurement, where the load power was kept constant, and the DC-AC conversion efficiency was determined by time average for 1 or 2-hour duration measurements after the initial variation. Figures 10(a) and 10(b) show the results for the step load and steady state measurements, respectively. In these figures, DC-AC conversion efficiencies are plotted with respect to the output power. In Figure 10(b), the output power dependence was obtained by setting the constant load value to be 5, 10, 20, 54, 90, 120, 154, and 179 W. For the step load measurement, the efficiencies at an out-put power of 54–179 W were approximately 86 and 81% for the SiC- and Si-based inverters, respectively. The difference in the efficiencies of the SiC- and Si-based inverters was approximately 5%, which decreased to approximately 3% under steady state measurement. This may have been attributed to the fact that the initial variation was more significant for the Si-based inverter than the SiC-based inverter. The decrease in efficiency at the low output power region also suggested the presence of non-load losses that were independent of the load power. The present electric circuit should be designed and wholly optimized for the SiC-FET in the further works [44-49].

6. LOSS ANALYSIS

The conduction loss of an inverter consists of contributions from a field-effect transistor (FET) (P_{fet}) and a body diode (BD) implemented between the source and drain (P_{bd}). Assuming a sinusoidal load current, P_{fet} and P_{bd} are expressed by [50]:

$$P_{\text{fet}} = R_{\text{ds}} \times I^2 \times \left(\frac{1}{8} + \frac{D}{3\pi} \cos\theta\right) \qquad (1)$$

and

$$P_{bd} = I \times V_f \times \left(\frac{1}{8} - \frac{D}{3\pi}\cos\theta\right), \tag{2}$$

where R_{ds}, V_f, I, D, and θ are the on-resistance of the FET, forward voltage of the BD, peak load current, modulation index, and power factor angle, respectively. The switching loss of the FET consists of contributions from the turn-on loss (P_{on}) and turn-off loss (P_{off}), which are expressed by [50]:

$$P_{on} = K_g \times E_{on} \times \frac{V}{V_t} \times \frac{I}{I_t} \times f_{sw} \tag{3}$$

and

$$P_{off} = K_g \times E_{off} \times \frac{V}{V_t} \times \frac{I}{I_t} \times f_{sw}, \tag{4}$$

where K_g, V, and f_{sw} are the correction factor, peak load voltage, and switching frequency, respectively. E_{on} and E_{off} are the turn-on and turn-off losses, respectively, for the test voltage V_t and current I_t. The reverse recovery loss of the BD is expressed by:

$$P_{rr} = \frac{1}{8}I_{rr} \times V \times t_{rr} \times f_{sw}, \tag{5}$$

where I_{rr} and t_{rr} are the reverse recovery current and reverse recovery time, respectively. The total loss P_{tot} of the DC-AC converter is then given by:

$$P_{tot} = P_{fet} + P_{bd} + P_{on} + P_{off} + P_{rr} + P_{nl}, \tag{6}$$

where P_{nl} is the non-load loss.

Table 4 summarizes the electrical characteristics for the SiC and Si MOSFETs operated at $T = 25$ °C. These parameters were taken from data sheets for SCT2120AF [42] and FQPF16N25C [41]. Although no data were available for E_{on} and E_{off} in [41], the switching times were approximately three times larger in FQPF16N25C than in SCT2120AF. For E_{on} and E_{off} in FQPF16N25C, the corresponding values in SCT2120AF were multiplied by three and used as an assumption. An increase in the forward current I across BD tended to result in an increase in I_{rr}. However, the short lifetime of minority carriers in SiC meant that I_{rr} was significantly suppressed in the SiC MOSFETs compared with in the Si MOSFETs. In this calculation, this effect was introduced by assuming that $I_{rr} = I/3$ for the SiC MOSFETs and $I_{rr} = 3I$ for the Si MOSFETs [51].

Table 4. Electrical characteristics of SiC and Si MOSFETs at $T = 25\ ^{\circ}C$

Property	Symbol	Unit	SiC	Si	Remarks
FET *on*-resistance	R_{ds}	Ω	0.11	0.22	Almost independent of I_d
FET turn-on loss	E_{on}	μJ	60	180	Measured at $V_t = 300$ V, $I_t = 10$ A
FET turn-off loss	E_{off}	μJ	40	120	Measured at $V_t = 300$ V, $I_t = 10$ A
BD forward voltage	V_f	V	2.15	0.73	Averaged for $I_f = 0.1$-1 A
BD reverse recovery time	t_{rr}	ns	33	260	Almost independent of I_f

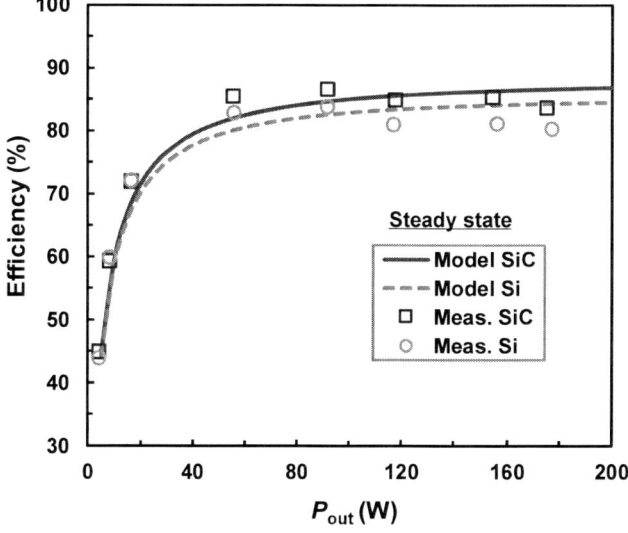

Figure 11. Modeled (lines) and measured (symbols) conversion efficiencies of SiC- and Si-based inverters with regard to the output power.

Figure 11 shows modeled DC-AC conversion efficiencies with regard to the output power ($V = 141.4$ V, $f_{sw} = 20$ kHz, $D = 1$, $\cos\theta = 1$, $K_g = 1$). In the calculations, the loss due to the front stage DC-DC converter was taken into account, and the DC-DC efficiency was assumed to be 90% for both inverters. The non-load loss was assumed to be 5 W. Measured steady state DC-AC conversion efficiencies are also shown. The calculations showed an approximate 2% improvement in the saturated efficiency of the SiC-based inverter compared with the Si-based inverter. This value was somewhat smaller than the measured value of approximately 3%. This discrepancy would arise from an efficiency degradation due to the self-heating effect occurring in the Si inverter, which was not included in the present analysis. As shown in Figure 6, measured efficiency of each inverter was gradually decreased by increasing the output power, and it deviated from the corresponding calculation. This tendency seemed to be more significant in the Si inverter compared with the SiC inverter. Judging from the lower thermal resistance of SiC MOSFETs (0.7 K W^{-1} for junction to case [42]) compared with Si MOSFETs (2.89 K W^{-1} for junction to case [41]), these behaviors were seemingly due to the self-heating effect of the switching devices. Enhanced self-heating effect in the Si MOSFETs should degrade their current drivability, and hence, lower the efficiency of the Si inverter. We presume this effect was responsible for the clear decrease in the efficiency of the Si inverter under high output power conditions ($P_{out} > 100$ W).

7. ESTIMATION OF POWER LOSS

Figures 12(a) and 12(b) show the dependence of total power loss on the output power of the SiC- and Si-based inverters, respectively. DC-DC converter losses and non-load losses are not included in Figure 12. The DC-AC converter loss in the SiC-based inverter was estimated to be one-third of that in the Si-based inverter.

Figures 13 and 14 shows the contribution of power loss elements (FET conduction loss P_{fet}, BD conduction loss P_{bd}, turn-on loss P_{on}, turn-off loss

P_{off}, and reverse recovery loss P_{rr}) in the SiC- and Si-based inverters occurring at various output powers in the range of 5~200 W. DC-DC converter losses and non-load losses are not shown.

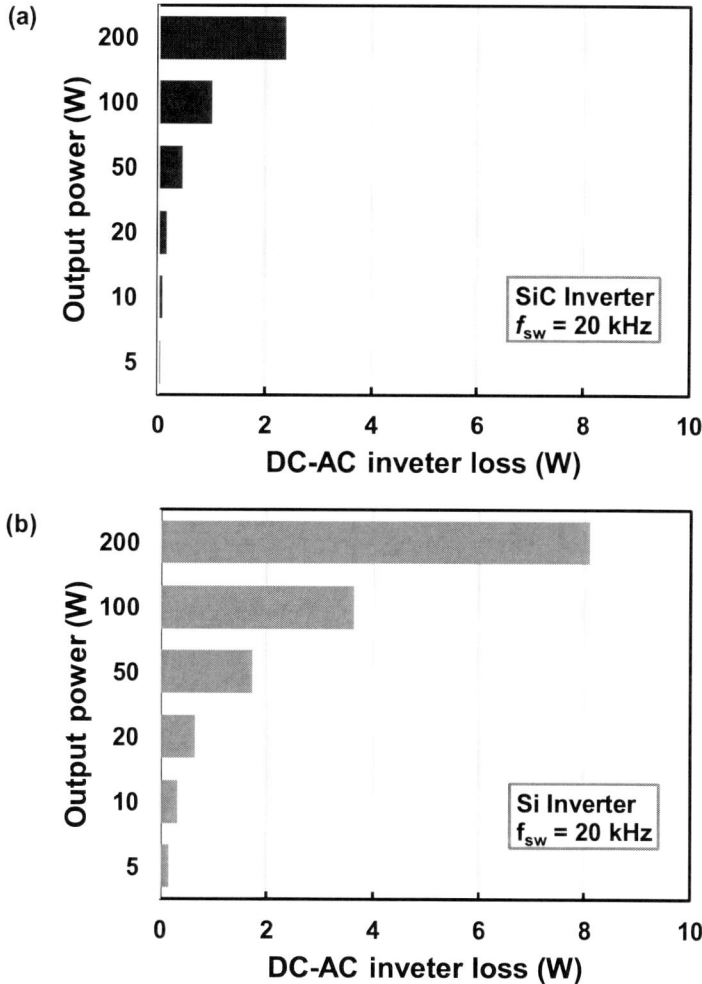

Figure 12. Dependence of DC-AC converter loss on the output power of the (a) SiC- and (b) Si-based inverters. DC-DC converter losses and non-load losses are not included.

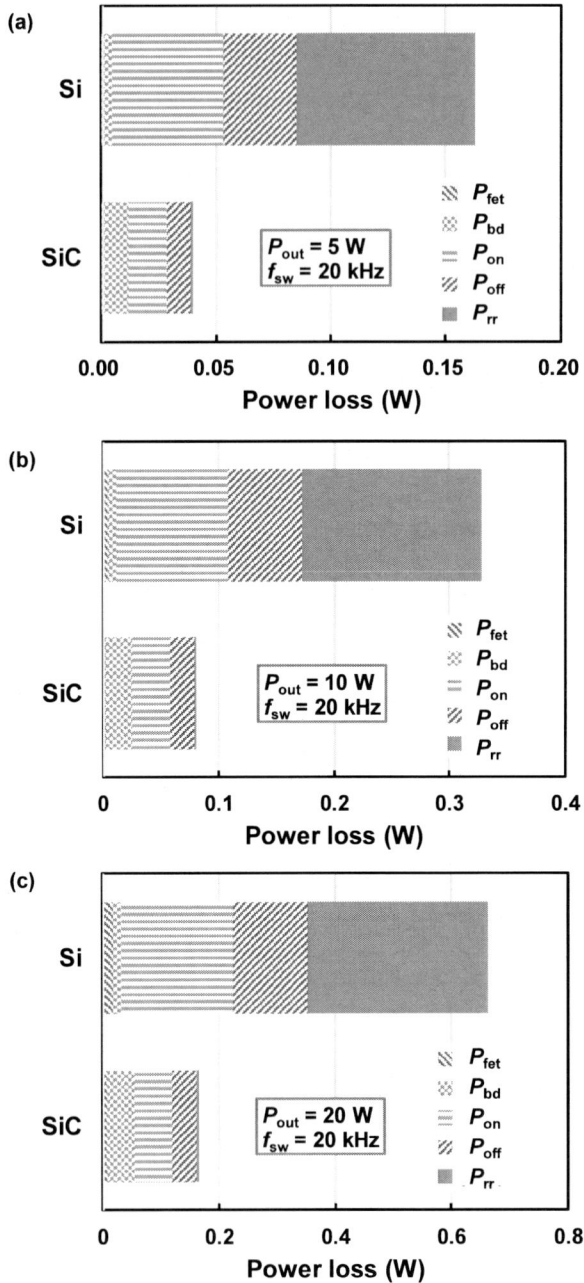

Figure 13. Contribution of power loss elements in SiC- and Si-based DC-AC converters occurring at output powers of (a) 5, (b) 10, and (c) 20.

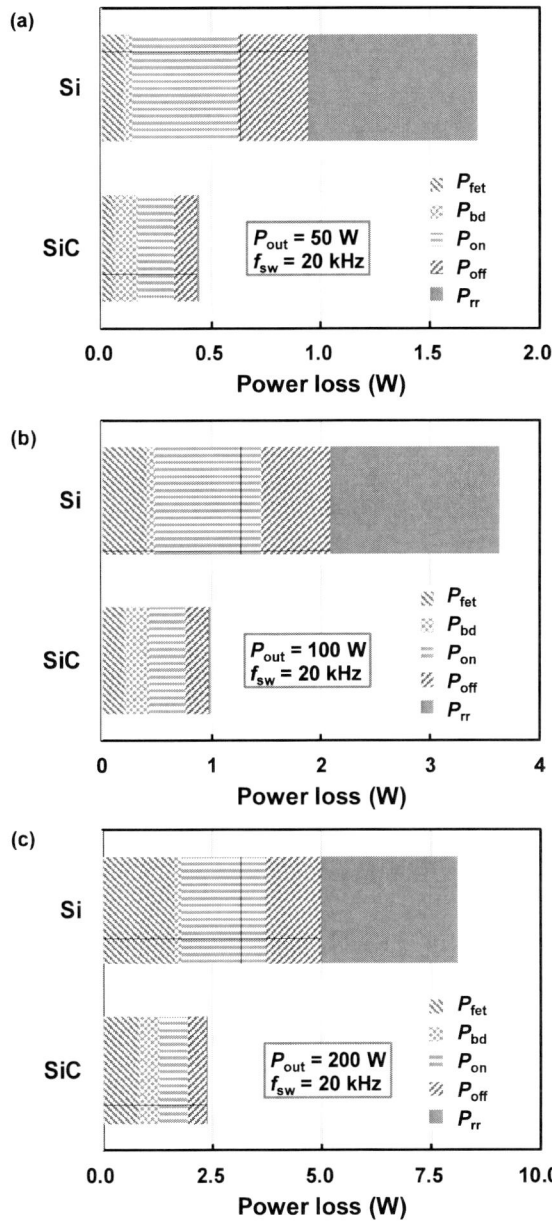

Figure 14. Contribution of power loss elements in SiC- and Si-based DC-AC converters occurring at output powers of (a) 50, (b) 100, and (c) 200 W.

The main power loss components in these inverters were the switching loss (P_{on} and P_{off}) and reverse recovery loss (P_{rr}). Decreases in these loss elements were responsible for the 2% improvement in efficiency of the SiC-based inverter. The contribution from conduction loss was more significant at higher output power. However, this calculation suggested that the loss improvement due to the small R_{ds} in a SiC MOSFET would be negated by the loss enhancement due to the large V_f of the body diode.

The main application of SiC power devices is multi-kW level converters. A 100-W level converter cannot fully benefit from the low *on*-resistance of SiC devices. However, the switching and reverse recovery losses were still improved in the SiC-converter, as shown in Figures 13 and 14, which has yielded the 3% improvement of the DC-AC conversion efficiency in the present 100-W class inverter.

The SiC inverter was prepared by replacing the Si MOSFETs in a conventional Si inverter with SiC MOSFETs, but the overall circuit remained optimized for Si MOSFETs. The switching frequency was also set to that of the conventional Si-inverter value. The switching frequency is related to the electric waveforms that are controlled by the filter circuit. Thus, simply replacing Si devices with SiC devices is far from an optimal design. In spite of this, the SiC inverter exhibited an efficiency superior to that of the conventional Si inverter. Optimizing the circuit is likely to further improve performance.

The current values decrease as the load increases, and it is believed that the effect of SiC-FET power devices become remarkable. When the output power of the inverters was below ~20 W, the DC-AC conversion efficiencies decrease to 70~40%, and the efficiencies of the inverter with Si-FET was higher than that with SiC-FET. This would be due to the decrease of effectiveness of R_{on} of SiC-FET for the low output power and/or to the originally existed resistances in the circuit. The present electric circuit system should be re-designed and optimized for the SiC-FET in the further works. Except for the present spherical Si solar cells, recent developed perovskite solar cells will also be applicable for such kinds of portable power generation systems with light weight [52-58].

CONCLUSION

The 100 W level power storage system consisting of an MPPT charge controller, spherical Si solar cells, a Li-ion battery, and SiC DC-AC converters was developed. The inverter used SiC MOSFETs as switching devices, and the SiC-based inverter exhibited a superior DC-AC conversion efficiency compared with that of the conventional Si-based inverter. Power loss analysis indicated that the *on*-resistance had little influence on the efficiency, and that a reduction in switching and recovery losses was re-sponsible for the higher efficiency of the SiC-based inverter. These results demonstrate the feasibility of SiC MOSFETs even for the 100 W level photovoltaic power generation systems.

ACKNOWLEDGMENTS

This work was supported by Super Cluster Program of Japan Science and Technology Agency (JST). We would like to aknowledge Koichi Hiramatsu, Yuya Ohishi, Yasuhiro Shirahata, Akio Shimono, Yoshikazu Takeda and Mikio Murozono for experimental help and supports.

REFERENCES

[1] Labedev, A. A.; Chelnokov, V. E. *Semiconductors* 1999, 33, 999-1001.
[2] Monroy, E.; Omnès, F.; Calle, F. *Semicond. Sci. Technol.* 2003, 18, R33-R51.
[3] Okumura, H. *Jpn. J. Appl. Phys.* 2006, 45, 7565-7586.
[4] Burger, B.; Kranzer, D.; Stalter, O. *Mater. Sci. Forum* 2008, 600-603, 1231–1234.

[5] Yaosuo, X.; Divya, K. C.; Griepentrog, G.; Liviu, M.; Suresh, S.; Manjrekar, M. *Energy Conversion Cong. Exposition (ECCE) IEEE* 2011, 2467.
[6] Schwarzer, U.; Buschhorn S.; Vogel, K. *Proc. PCIM Europe, IEEE* 2014, 787–794.
[7] Thooyamani, K. P.; Khanaa V.; Udayakumar, R. *Mid. East J. Sci. Res.* 2014, 20, 2059–2064.
[8] Oku, T. *Solar Cells and Energy Materials*; De Gruyter; Berlin, Germany, 2016.
[9] Burger, B.; Kranzer, D.; Stalter, O.; *5th Int. Conf. Integrated Power Systems (CIPS)* 2008, 1–5.
[10] Kim, T.; Jang, M.; Agelidis, V. G. *IEEE ECCE Asia Downunder* (ECCE Asia), Melbourne, Australia, 3-6 June 2013, 555-559.
[11] Chinthavali, M.; Zhang, H.; Tolbert, L. M.; Ozpineci, B. *Power Electronics Conf.* 2009 (COBEP '09), 71–79.
[12] Hensel, A.; Wilhelm, C.; Kranzer, D. Proc. *14th Euro. Conf. Power Electronics and Applications* 2011 (EPE 2011) 1–7.
[13] Yamane, A.; Koyanagi, K.; Kozako, M.; Fuji, K.; Hikita, M. *IEEE 10th Int. Conf. Power Electronics and Drive Systems* 2013 (PEDS 2013), 1029–1032.
[14] Li, L.; Li, C.; Y. Cao, Y.; Wang, F. *IEEJ Trans. Electrical and Electronic Eng.* 2013, 8, 515–521.
[15] De, D.; Castellazzi, A.; Solomon, A.; Trentin, A.; Minami, M.; Hikihara, T. *15th Euro. Conf. Power Electronics and Applications* 2013 (EPE 2013), 1–10.
[16] Nashida, N.; Hinata, Y.; Horio, M.; Yamada, R.; Ikeda, Y. *IEEE 26th Int. Symp. Power Semiconductor Devices & IC's* 2014 (ISPSD 2014) 342–345.
[17] Sintamarean, C.; Eni, E.; Blaabjerg, F.; Teodorescu, R.; Wang, H. *Power Electronics Conf.* 2014, 1912–1919.
[18] Zhang, H.; Tolbert, L. M.; Ozpineci, B.; Chinthavali, M. S. *Industry Applications Conf.* 4, 2005, 2630–2634.

[19] Matsuomoto, T.; Oku, T.; Hiramatsu, K.; Yasuda, M.; Shirahata, Y.; Shimono, A.; Takeda, Y.; M. Murozono, *AIP Conf. Proc.* 2016, 1709, 020023-1–6.
[20] Oku, T. *Solar Cells and Energy Materials*. De Gruyter, Berlin, Germany, 2017.
[21] Oku, T.; Matsumoto, T.; Hiramatsu, K.; Yasuda, M.; Shimono, A.; Takeda, Y.; Murozono, M. *AIP Conf. Proc.* 2015, 1649, 79–83.
[22] Maruyama, T.; Minami, H. *Sol. Energy Mater. Sol. Cells* 2003, 79, 113–124.
[23] Gharghi, M.; Sivoththaman, S. *J. Electron. Mater.* 2008, 37, 1657–1664.
[24] Liu, Z.; Masuda, A.; Kondo, M. *J. Cryst. Growth* 2009, 311, 4116–4122.
[25] Okamoto, C.; Minemoto, T.; Murozono, M.; Takakura, H.; Hamakawa, Y. *Jpn. J. Appl. Phys.* 2005, 44, 7805–7808.
[26] Liu, Z.; Nagai, T.; Masuda, A.; Kondo, M.; Sakai, K.; Asai, K. *J. Appl. Phys.* 2007, 101, 093505.
[27] Gharghi, M.; Bai, H.; Stevens, G.; Sivoththaman, S. *IEEE Trans. Electron Devices* 2006, 53, 1355–1363.
[28] Minemoto T.; Takakura, H. *Jpn. J. Appl. Phys.* 2007, 46, 4016–4020.
[29] Ono, Y.; Oku, T.; Akiyama, T.; Kanamori, Y.; Ohnishi, Y.; Ohtani, Y.; Murozono, M. *J. Phys. Conf. Series.* 2012, 352, 012023-1–5.
[30] Oku, T.; Kanayama, M.; Akiyama, T.; Kanamori, Y.; Murozono, M. *Phys. Stat. Solidi C* 2013, 10, 1840–1843.
[31] Oku, T.; Kanayama, M.; Ono, Y.; Akiyama, T.; Kanamori, Y.; Murozono, M. *Jpn. J. Appl. Phys.* 2014, 53, 05FJ03-1–7.
[32] Shirahata, Y; Zhang, B.; Oku, T.; Kanamori, Y.; Murozono, M. *Mater. Trans.* 2016, 57, 1082–1087.
[33] Shirahata, Y.; Oku, T.; Kanamori, Y.; Murozono, M. *J. Ceram. Soc. Jpn.* 2017, 125, 145–149.
[34] Shirahata, Y.; Oku, T.; Kanamori, Y.; Murozono, M. *AIP Conf. Proc.* 2016, 1709, 020021-1–8.
[35] Shirahata, Y.; Oku, T.; Kanamori, Y.; Murozono, M. *AIP Conf. Proc.* 2017, 1807, 020019-1–7.

[36] Nakamura, T.; Akashi, Y.; Murozono, M. *Japan Patent* 2010, 2010-150106.
[37] Kanamori, Y.; Akashi, Y.; Murozono, M. *Japan Patent* 2012, 2012-126592.
[38] Murozono, M.; Ohshima, Y.; Hibino, T. *Japan Patent* 2012, 2012-17234.
[39] Ando, Y.; Shirahata, Y.; Oku, T.; Matsumoto, T.; Ohishi, Y.; Yasuda, M.; Shimono, A.; Takeda, Y.; Murozono, M. *J. Power Energy Eng.* 2017, 5, 30–40.
[40] Oku, T.; Matsumoto, T.; Hiramatsu, K.; Yasuda, M.; Ohishi, Y.; Shimono, A.; Takeda, Y.; Murozono, M. *AIP Conf. Proc.* 2016, 1709, 020024-1–10.
[41] Fairchild Semiconductor, Datasheet of FQPF16N25C—N-Channel QFET MOSFET, 2013. Available online, https://www.onsemi.jp/pub/Collateral/FQP16N25CJP-D.pdf.
[42] ROHM Semiconductor, Datasheet of SCT3060AL – N-Channel SiC Power MOSFET, 2016. Available online, http://www.rohm.com/web/global/datasheet/SCT3060AL/sct3060al-e.
[43] Beijing EPsolar Technology, Tracer2215BN, Available online, https://www.epsolarpv.com/upload/cert/file/1811/Tracer-BN-SMS-EL-V1.2.pdf.
[44] Ando, Y.; Oku, T.; Yasuda, M.; Shirahata, Y.; Ushijima, K.; Murozono, M. *AIP Conf. Proc.* 2017, 1807, 020020-1–9.
[45] Oku, T.; Ando, Y.; Yasuda, M.; Shirahata, Y.; Ushijima, K.; Murozono, M. *Adv. Energy Power* 2017, 5, 7–12.
[46] Ando, Y.; Oku, T.; Yasuda, Y.; Ushijima, K.; Murozono, M. *Technologies* 2017, 5, 18-1–11.
[47] Ando, Y.; Oku, T.; Yasuda, M.; Shirahata, Y.; Ushijima, K.; Murozono, M. *Solar Energy* 2017, 141, 228–235.
[48] Ando, Y.; Oku, T.; Yasuda, M.; Ushijima, K.; Matsuo, H.; Murozono, M. *AIP Conf. Proc.* 2018, 1929, 020002-1–9.
[49] Ando, Y.; Oku, T.; Yasuda, M.; Ushijima, K.: Matsuo, H.; Murozono, M. *Heliyon* 2020, 6, e03094-1–10.

[50] Pattnaik, S.K.; Mahapatra, K.K. *Proc Inter. Multiconf. Engineers and Computer Scientists 2010* (IMECS 2010), Hong Kong, 17-19 March 2010, 1401-1406.

[51] ROHM Semiconductor, (2014) SiC Power Devices and Modules—Application Note. Available online, http://rohmfs.rohm.com/en/products/databook/applinote/discrete/sic/common/sic_appli-e.pdf.

[52] Oku, T.; Ohishi, Y.; Ueoka, N. *RSC Advances* 2018, 8, 10389–10395.

[53] Ueoka, N.: Oku, T. *ACS Appl. Mater. Interfaces* 2018, 10, 44443–44451.

[54] Oku, T.; Nomura, J.; Suzuki, A.; Tanaka, H.; Fukunishi, S.; Minami, S.; Tsukada, S. *Int. J. Photoenergy* 2018, 2018, 8654963-1–7.

[55] Machiba, H.; Oku, T.; Kishimoto, T.; Ueoka, N.; Suzuki, A. *Chem. Phys. Lett.* 2019, 730, 117–123.

[56] Kishimoto, T.; Suzuki, A.; Ueoka, N.; Oku, T. *J. Ceram. Soc. Jpn.* 2019, 127, 491–497.

[57] Taguchi, M.; Suzuki, A.; Oku, T.; Ueoka, N.; Minami, S.; Okita, M. *Chem. Phys. Lett.* 2019, 737, 136822-1–7.

[58] Ueoka, N.; Oku, T.; Suzuki, A. *RSC Advances* 2019, 9, 24231–24240.

In: Maximum Power Point Tracking
Editor: Maurice Hébert

ISBN: 978-1-53618-164-7
© 2020 Nova Science Publishers, Inc.

Chapter 4

MAXIMUM POWER POINT TRACKING: A REVIEW OF THE CONSIDERATIONS FOR LARGE SCALE AND SMALL SCALE PHOTOVOLTAIC INSTALLATIONS

Sarah Lyden[*]
School of Engineering, University of Tasmania, Hobart, TAS, Australia

ABSTRACT

A large variety of Maximum Power Point Tracking (MPPT) techniques have been proposed in the literature ranging from simple approximation techniques to those based on computationally complex algorithms. In general, the development of complex techniques is aligned with the goals of improving MPPT performance under non-uniform environmental conditions which could occur due to cell damage or shading between cells such as what would be typically seen in a residential environment. Photovoltaic (PV) systems are frequently

---

[*] Corresponding Author's E-mail: Sarah.Lyden@utas.edu.au

deployed in residential and utility scale implementations and the requirements of these installations in terms of control can be quite distinct. This chapter will review the considerations of MPPT in large utility scale PV systems and small-scale residential PV systems identifying common requirements and differences in the appropriateness of MPPT techniques applied. A set of characteristics of large scale and small-scale PV installations will be proposed, and criteria defined to evaluate the suitability of a technique for application in a large scale or small-scale system. Common and emerging MPPT techniques will be evaluated against these criteria. The main contribution of the chapter is highlighting the considerations in selecting an appropriate MPPT technique based on the scale of the PV system.

Keywords: Maximum Power Point Tracking, photovoltaic, utility-scale, small-scale

INTRODUCTION

Maximum Power Point Tracking (MPPT) is a widely researched area in the optimisation of Photovoltaic (PV) systems. To date, more than 30 distinct MPPT techniques have been proposed with varying levels of complexity, cost and dependence on system parameters [1]. Prior reviews of MPPT techniques have considered the capabilitiy of these techniques to identify local and global maxima, the cost involved in the techniques, and complexity suitability to apply to other PV systems [1–3]. The review presented in this chapter is different as it approaches consideration of MPPT technique and their capabilities from the perspective of identifying the different requirements of large scale (utility) and small scale (residential) PV systems. In particular the relevance of advanced MPPT technqiues for each type of system is described.

Renewable energy technology is being increasingly adopted around the world to address concerns of the environmental impact of conventional sources of energy. Most renewable energy sources remain out of reach of consumers as these sources, including wind turbines and hydro power, require suitable geographical conditions and are typically scaled beyond what a residential consumer would require. Technology such as PV cells however have become accessible to residential consumers as these energy sources can be easily deployed as distributed generation. Additonally, PV is a unique source as it is scalable and can be deployed in both small scale residential settings as well as utility scale PV plants. In either case, extracting the maximum power from the PV system is important as these systems typically still have a high cost of energy, as despite recent cost declines the PV efficiency is still rather low (less than 25% in most commercial PV cells e.g., [4, 5], although current experimental cells are reaching efficiencies of 47.1% [6]).

The considerations involved in extracting the maximum power from both utility scale and residential scale PV systems are outlined in this chapter. In particular this chapter seeks to classify existing MPPT techniques by their appropriateness for application to residential scale or utility scale PV systems.

In a residential scale PV system there is likely to be shading from objects in the environment such as trees, other houses, and power poles which may cause shading on the system throughout the day. There is also the possibility of inter-row shading depending on how the panels are oriented on the roof and the spacing between them [7]. Often these shading objects are fixed in the environment and cannot be easily removed, so the capability of a residential based MPPT to be able to identify global maxima becomes increasingly important to ensure that, proportional to system size, a large amount of PV energy is captured.

In this paper, a utility scale PV plant is defined as one requiring a large geographical area and typically rated at more than 100 kW. The area for a utility scale PV installation is typically chosen away from obstacles that could result in partial shading of the system, however because of the large geographical areas the system may cover, non-uniform conditions could still be experienced. Additionally, these systems may be plagued by small scale mismatch between the cells or modules due to manufacturing tolerance or cell damage and aging [8].

In order to assess each of these environments and the unique requirements guiding the choice of most suitable MPPT technique it is necessary to consider factors such as the significance of cloud transients on

PV systems of all sizes, what size of PV system is considered to be exposed to uniform weather conditions and if the advantages posed by distributed MPPT (DMPPT) [9–11] are applicable in all systems sizes. This chapter utilises the literature related to these key factors in developing criteria to assess the suitability of particular MPPT techniques for each application. In doing so this chapter seeks to answer the question are the well established MPPT methods suitable for both applications or is a different approach needed when considering residential scale compared with utility scale systems?

In order to assess the extent to which applications or classification of a scale for a particular technique is highlighted in the literature, IEEExplore was used to do an indicative search. When searching with terms "MPPT" and "PV," 4845 results appear on IEEExplore. When refining this further to be:

- "MPPT," "PV," "small-scale" or "residential" 122 results are found
- "MPPT," "PV," "large-scale" or "utility" 303 results are found

Further investigation of these results highlights that microgrid applications often appear in either study in addition to hybrid systems which are considered outside the scope of this chapter.

When considering MPPT with small scale or residential, the majority of papers investigate the use of standard techniques such as Perturb and Observe (P&O) [12–14], Incremental Conductance (IncCond) [15] and their extensions [16–18], or consider topology-based approaches or the design of grid connected inverters [19–21].

When considering MPPT with large scale or utility scale, a large number of the papers found rely on using conventional techniques such as P&O while the focus of the paper is typically on the interaction of the system with the utility grid [22–24], which often requires the maximum power of the PV system to be curtailed, for instance to avoid exceeding ramp rate requirements [25–28]. In these papers MPPT appears to be just a secondary concern and common techniques are often applied.

Given that other factors such as the ramp rate control and grid interconnection requirements must be met when connecting a utility scale PV system to the grid, these requirements may often override MPPT. In this case the necessity of extracting the full potential from the system using an advanced technique may not be worthwhile when a conventional technique can adequately perform the same function.

Outside of those papers captured by this literature search very few papers specifically mention the size of system their technique is most appropriate to be implemented on. In many cases a degree of scalability appears to be assumed by applying methods validated on small scale systems into increasingly larger systems. This leads to a further question if a technique has been designed to deal with non-uniform conditions of a particular type (for example solid shading from objects in an environment) is the same technique equally applicable for different cases of shading (for example that caused by uneven cloud cover across a system)? Essentially, are shading phenomenon caused by weather variations in the form of cloud movement different in the effects on PV systems when compared with shading from solid objects in the local environment of the system? This will be specifically considered in this chapter.

This introduction has provided several key questions which will be explored throughout the remainder of this chapter in particular:

1) To what extent can cloud transients affect both small scale and large-scale systems?
2) In utility scale systems, where power often needs to be curtailed to meet grid requirements, are advanced techniques really needed?
3) Does shading from clouds and shading from solid objects in an environment produce significantly different MPPT needs?
4) Is the development of MPPT scalable?
5) Are the well established MPPT methods suitable for both applications or is a different approach needed when considering residential scale compared with utility scale systems?

The remainder of this chapter is organised as follows, a review of MPPT methods is provided, characteristics of the different scale systems are defined, criteria for evaluating the performance of MPPT in different scales are defined and a table showcasing common and emerging techniques is provided. Finally, conclusions highlighting the considerations found from this chapter in selecting an appropriate MPPT technique based on the scale of the system are provided.

MAXIMUM POWER POINT TRACKING TECHNIQUES

In general MPPT techniques can be classified as conventional, advanced or topology based techniques. Conventional techniques are those which work well under uniform conditions and typically have a low complexity. Some key examples include P&O [29], and fractional open circuit voltage [30]. Advanced techniques are those which are typically designed to address the issue of partial shading and multiple peaks in the power-voltage (P-V) characteristic. These techniques often emulate biological processes or utilise advanced computational techniques to distinguish between local and global maxima such as Particle Swarm Optimisation (PSO) [31] and Simulated Annealing (SA) [32]. Topology based approaches include the granularity of where converters with associated MPPT are placed in the system including distributed MPPT (DMPPT) approaches [33] and those approaches where the modules can either be reconfigured or are not connected in the standard series-parallel string formation [34].

MPPT is applied in the DC-DC interface (or in the DC-AC interface in single stage implementation). Basically the role of this converter is to regulate the changing PV voltage into the constant DC link voltage or grid voltage. As the output side of the DC-DC converter is fixed, the input voltage to the converter will change with the change in duty cycle enacted by the MPPT controller. A typical two stage PV system is shown in Figure 1.

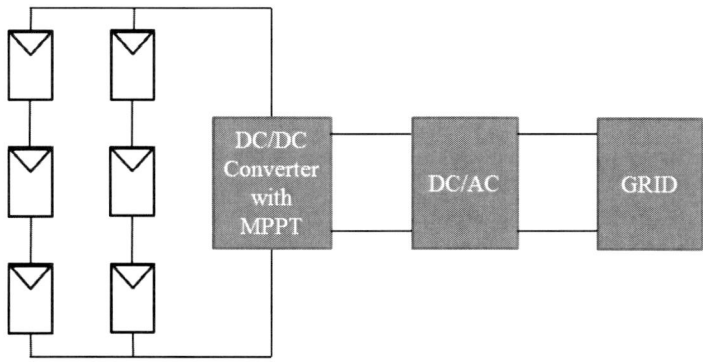

Figure 1. Two stage converters with MPPT implemented in DC-DC converter.

Conventional MPPT Techniques

Conventional MPPT techniques are those designed to work well in uniform environmental conditions and provide tracking usually with some trade-off between speed and accuracy due to the simple approach of the techniques. One of the main conventional MPPT techniques which has received much attention in the literature is the P&O method [17, 35, 36]. This method applies a perturbation and observes the change in power to indicate if a step should be taken in the same direction or if the search direction should be reversed to locate the maximum power point. This technique works on the basis that as the perturbations are approaching the MPP the power will be getting larger, while when passing the MPP or stepping away the power will be becoming smaller. The P&O technique has a trade-off between speed and accuracy – a large step size leading to a quick search will result in power losses as the technique oscillates around the optimal point. Additionally, the P&O method has no capability to distinguish if a located maxima is a local or global maxima which could lead to considerable lost power potential in a system. To overcome some of these limitations variable step size MPPT methods and combining the method in a two stage implementation to improve outcomes under non-uniform conditions are common approaches [17, 37].

A similar technique to P&O is the IncCond method [15] which also takes a perturbation based approach however, seeks the point where the instantaneous conductance is equal to the incremental conductance. IncCond typically has better performance than P&O in terms of robustness to measurement noise [29].

Other conventional techniques include the fractional open circuit voltage method where an approximate relationship between the location of the MPP voltage is defined based on the measured open circuit voltage [3, 30]. Limitations with this technique are that it relies on an approximation which may degrade with time and, in temporarily open circuiting the system to take a measurement no power will be produced.

There are a wide range of other conventional techniques proposed in the literature, for this study the most common of these are focussed on. A more extensive literature review of MPPT methods including the wide range of conventional techniques can be found in [1–3].

Advanced MPPT Techniques

Advanced MPPT techniques are those that have added complexity when compared with the conventional techniques and are usually designed to be able to identify a global maximum under non-uniform environmental conditions.

Some key examples in the literature include PSO [31, 38], grey wolf optimization (GWO) [39], firefly algorithm (FA) [40], SA [32], and Bayesian optimization (BO) [41, 42].

PSO is based on the flocking behaviours of birds and schooling of fish in order to collectively solve a problem. A number of particles are used to move towards the global optimum by considering the global best position and best position of each of the particles [31].

The GWO MPPT technique is based on the leadership hierarchy of wolves in hunting prey [39]. This technique is able to quickly identify a global peak and relies on fewer parameters requiring adjustment than some of the other common evolutionary approaches [39].

FA is based on the flashing behaviour of fireflies in attracting mates and research has shown that the PSO algorithm is actually a special case of the generalized FA [40]. The FA, like other techniques, will require a reinitialization condition in order to detect when a change in conditions occurs. In [40] this is implemented as monitoring the change in output power, however as shown in [43], techniques such as PSO and FA may require a more advanced reinitialization condition to accurately detect all cases requiring a new global search.

SA is a MPPT technique based on the emulating the annealing process used in metals to find a particular energy state. SA has few parameters and minimal complexity, however requires a reinitialization condition when a change in shading conditions occurs to operate effectively [44].

BO is a recently proposed approach for locating a global MPP. This technique relies on viewing the unknown objective function as a random function characterized by a prior distribution. This is then combined with a likelihood distribution function to produce a posterior function, which then informs an acquisition function which guides the choice of next sample point leading to convergence. While the technique is very effective it does rely on good selection of parameters to operate well [42].

These advanced techniques usually have increased computational complexity when compared to the conventional techniques, but in general have good capability in identifying global peaks in the P-V characteristic. Often these techniques have parameters which are tuned for a particular system or application thus limiting their universal applicability to all systems.

Topology-Based MPPT Techniques

Topology based MPPT techniques usually combine a conventional MPPT method with a change in the configuration of the panels or positioning of the DC-DC converter in order to increase energy capture under complex environmental conditions. One key approach to using the topology is DMPPT [33]. In this approach, smaller sections of the system, made up of strings or even single modules, have their own dedicated DC-DC converter. This distributed architecture means that the smaller sections can achieve their own best operating point and are not constrained by other modules in the system which may be experiencing shading. This architecture also has the advantage of using smaller more readily available converters and having redundancy as the system is not reliant on a single centralised DC-DC converter for its operation.

Other topology based approaches include differential power processing (DPP) where the converters are placed between PV modules [45]. Where there is a mismatch between the modules the converter will only process the power difference. This enables the converters to be sized significantly smaller.

Additionally, rather than connecting modules in the standard series-parallel configuration other options such as bridge-linked (BL) and total cross tied (TCT) configurations can lead to improved energy capture even with conventional MPPT techniques applied [46, 47].

METHOD

In order to perform a comparison of the use of MPPT methods for both small-scale residential and large-scale utility PV systems it is necessary to first define these terms. For the purposes of this chapter:

- Small scale (residential) - a PV system of less than 10 kW which is installed on the roof top of a residential building.
- Large scale (utility) – a PV system of more than 100 kW which is installed by a utility company or private contractor usually on an expanse of land rather than on a structure.

While it may be possible for utility scale system to also be installed in a more complex residential environment (such as on the roof top of a car park building for example), this would represent a hybrid case based on the considerations outlined in Table 1. Table 1 presents the key characteristics of small-scale (residential) and large-scale (utility) PV systems.

Table 1. Key characteristics of small scale and large scale PV systems

	Small scale (residential)	Large scale (utility)
Size	Small geographical area covered, irradiance across the system should be constant with environmental conditions however will be affected by transition of shadow edges across the system due to clouds	Large geographical area covered Potential for different irradiance to be experienced across different parts of the system due to the passage of clouds, affected by transition of shadow edge across system due to clouds
Obstacles in environment which may cause shading	Likely due to non-ideal environment (trees, power poles, roof artefacts etc.)	Less likely due to the design and placement of the system Inter-row shading possible if insufficient spacing between modules allowed
System configuration	Typically, single or multiple strings of modules with centralised power electronics, although module level converters can be used.	Typically integrates some form of distributed converters for reliability, however centralised converters can also be used.
Reliability/redundancy	Without module level converters dependent on centralised converter	Typically has redundancy if distributed power converters are used
Characteristic requirements	Exports excess produced power to the utility grid or charges a local battery	Need to adhere to utility defined ramp rate specifications
Goals	Consumer driven goals – maximise energy production, offset their own costs	Utility operator driven goals – support power grid, provide renewable energy sources

The characteristics table suggests that consumers are most motivated by getting a return on their investment in a PV system by producing the maximum amount of power at any point in time throughout the day and either using this for their own use or exporting to the grid. For the utility operated PV system, while maximising return for investors is also an important consideration, due to the size of this plant it also has a role in supporting the grid. That means that the utility PV system must implement ramp rate control such that the transients in power due to changing environmental conditions do not occur too quickly to not be able to be offset by other generation sources. This means that large scale systems will not always be operating in an MPPT state and in some instances may need to operate below the MPP to ensure there is ramp rate potential (or to curtail power available) if there is no energy storage included in the system [48]. On this basis, conventional MPPT with appropriate designed strings and use of DMPPT can lead to a similar outcome to advanced techniques with lower cost and complexity.

Based on these characteristics of the different types of systems under consideration in this chapter a set of requirements for MPPT in different situations can be developed.

For a residential setting:

1) Maximize power extraction while minimizing costs
2) Capable of handling shading from objects in environment

For utility scale:

3) Capable of being integrated with ramp rate control and other necessary grid control schemes thereby often operating below the MPP
4) Capable of handling cloud transients

The defined criteria for MPPT in residential and utility scale PV systems given above are used to evaluate several common MPPT approaches in Table 2. Conventional, advanced and topology-based approaches are considered in terms of how they deliver these key requirements in different environments.

DISCUSSION

The key characteristics of PV systems of residential and utility scale have been defined and utilised to provide key criteria for MPPT in each application. Several key MPPT approaches ranging from conventional techniques to topology approaches have been compared using these criteria.

Table 2. Comparison of techniques utilizing defined criteria

	Residential		Utility	
	Criteria 1	*Criteria 2*	*Criteria 3*	*Criteria 4*
Conventional techniques				
P&O [14]	Low cost, good at maximizing power under uniform conditions	Unable to distinguish global peaks under complex shading	Yes	Unable to distinguish global peaks under non-uniform shading, configuration will influence
Fractional Open circuit Voltage [30]	Low cost, performance degrades with time, zero power when sampling	Unable to distinguish global peaks under complex shading	Yes	Unable to distinguish global peaks under non-uniform shading
Advanced techniques				
PSO [31]	Moderate computational cost, good performance	Effective at locating global maxima under complex shading, performance dependent on initial positions	If power curtailed, then high computational cost algorithm to then operate below optimal power point	Yes, although performance is dependent on initial particle positions
GWO [39]	Moderate computational cost, good performance	Effective at locating global maxima under complex shading	If power curtailed, then high computational cost algorithm to then operate below optimal power point	Yes
FA [40]	Low-moderate computational cost, good performance	Effective at locating global maxima under complex shading, needs an effective reinitialization condition	If power curtailed, then high computational cost algorithm to then operate below optimal power point	Yes, although suitable reinitialization condition needed to detect when the clouds are moving
Advanced techniques				
SA [32, 49]	Low-moderate computational cost, good performance	Effective at locating global maxima under complex shading, needs reinitialization	If power curtailed, then relatively high computational cost algorithm	Yes, however suitable reinitialization condition needed for clouds

Table 2. (Continued)

	Residential		Utility	
	Criteria 1	Criteria 2	Criteria 3	Criteria 4
Advanced techniques				
BO [42]	Moderate computational cost, good performance	Highly effective at locating global maxima with suitably defined parameters	If power curtailed, then relatively high computational cost algorithm to then operate below optimal power point	Yes, effective as long as suitable parameters used in the method implementation
Topology based techniques				
DMPPT [11]	Potentially higher costs due to additional converters Maximises power extraction while using conventional techniques	Depending on the granularity and choice of MPPT technique may not be able to extract absolute maximum power from the system	Yes, grid integration measures can be implemented from a centralised converter	Yes, if strings suitable sized and aligned appropriately for the cloud transient
DPP [8]	Low cost as converters only process power difference	Unable to address mismatch within modules unless coupled with advanced MPPT method	Yes, grid integration measures can be implemented from a centralised converter	Yes, although unable to address internal mismatch to the cell unless coupled with advanced MPPT method
TCT [46]	Higher cost and complexity	Needs to be coupled with a technique capable of distinguishing between local and global maxima	Yes, grid integration measures can be implemented from a centralised converter	Yes, however requires a MPPT method capable of locating global maxima
BL [46]	Moderate cost and complexity	Needs to be coupled with a technique capable of distinguishing between local and global maxima	Yes, grid integration measures can be implemented from a centralised converter	Yes, however requires a MPPT method capable of locating global maxima

The key findings from this table will be discussed with reference to key findings from the literature informing how PV systems are likely to be affected by the transition of clouds and shadow from other objects. Finally, these results and this discussion will lead to consideration of the questions proposed in the introduction section to this chapter.

Work completed by Lappalainen and Valkealahti has shown significant results in terms of the impact of cloud transitions on irradiance and the flow on effect for PV. Some of their key findings include, if the shadow strength is below 40% it does not significantly affect the power produced [50]. They showed that the apparent shadow edge speed has a more significant impact on power generation than the speed of movement. This apparent shadow edge speed, which is the speed perpendicular to the shadow from a cloud, has an average speed of around 9 m/s and the length of irradiance transitions caused by the edge of a moving cloud is around 100 m [51]. Further work showed that the duration of shading periods on PV systems due to clouds varies from seconds up to 1.5 hours [52]. These irradiance transitions due to the shadow caused by clouds affect both large scale and small scale systems, however the total amount of time that irradiance transitions affect a PV system was found to be around a half hour each day [51] leading to mismatch losses of less than 1% of overall production in large scale systems [47].

Studies of these effects on the sizing of a system and the configuration showed that shorter string lengths are better in terms of minimizing the mismatch due to cloud transitions. Shading caused by moving clouds is not a severe change in irradiance on a system and therefore leads to only minor differences between the irradiance experienced by adjacent PV modules [47].

Sharp shadows, such as those resulting from objects in the environment around a PV module lead to a significantly higher mismatch loss than the shadow caused by clouds [47]. This is in line with observations in this chapter that, in an environment where there are likely to be fixed objects which will shade the panels (typical of a residential environment), then the application of advanced MPPT techniques capable of distinguishing between local and global maxima becomes more relevant.

Prior research suggests that the shading caused by objects in an environment is more likely to lead to severe shading conditions in PV systems than that caused by the passage of clouds across a system [47, 53]. If a large-scale PV system is designed appropriately with strings that are reasonably designed, then the impact of this shadow is typically minimised. However, these systems may benefit from something like forecasting in order to interface the MPPT with ramp rate control and other scheduling parameters for a large-scale grid connected plant. In general, in a system relying on DMPPT, conventional techniques could be applied to good effect as the amount of inter-string shading and variation can be minimised by effective system design.

In a residential setting it may not be practical to implement module level converters and the type of shading may move more slowly across the system and have a stronger effect, therefore the capability to implement Global (GMPPT) becomes more relevant if the goal is to maximize power produced. Additionally, though, it is expected that shading from objects in the environment has a significant effect on the P-V characteristics, these objects are often fixed in the environment and will only cause shading over certain periods of the day, suggesting that GMPPT combined with conventional techniques for when conditions are fairly standard could be a way to optimise system performance.

Appropriately placed DMPPT may be best for large scale systems with conventional techniques as it also enables redundancy, improves reliability, but may add to costs [54, 55] and because the non-uniform conditions may change a lot quicker than from static shading this would ensure better overall energy capture. Compared with, for instance, the residential setting where during the times of day when static shading is most likely to be a problem having a GMPPT routine in place would be advantageous to overcome the limitations of some shadow.

Based on the analysis and literature review presented in this chapter the key questions stated at the beginning can now be addressed.

To What Extent Can Cloud Transients Affect Both Small Scale and Large-Scale Systems?

The literature has shown that the cloud shadow edge effects both small-scale and large-scale systems however large-scale systems are more likely to have more significant levels of non-uniform irradiance across them than smaller systems. The time when cloud shadow transient edges are likely to affect PV systems was found to be quite small in prior studies [51].

In Utility Scale Systems Where Power Often Needs to Be Curtailed to Meet Grid Requirements Are Advanced Techniques Really Needed?

Given that power may be curtailed in a large-scale PV system, it appears as though DMPPT or topology-based approaches coupled with conventional, simple MPPT methods may be the most appropriate solution. While an advanced technique may lead to an increase in energy capture under certain circumstances, it may not be worth the additional implementation cost and computational burden of the technique.

Does Shading from Clouds and Shading from Solid Objects in an Environment Produce Significantly Different MPPT Needs?

Yes, the literature has suggested that shading from cloud and shading from solid objects affect the P-V characteristics differently [47, 53]. This further suggests that the MPPT concerns under these different conditions are different.

Is the Development of MPPT Scalable?

While in many cases in the literature the scalability of the MPPT methods appears to be assumed, based on literature results without some kind of additional technology such as the use of shorter strings [47], changes in topology or DMPPT these techniques are not directly scalable.

Are the Well Established MPPT Methods Suitable for Both Applications or Is a Different Approach Needed When Considering Residential Scale Compared with Utility Scale Systems?

A key outcome of this chapter is the fact that complementary techniques are needed in utility scale PV systems such as integrated DMPPT with a conventional or advanced technique. In these approaches the well-established methods work well, although the requirement or benefit of a substantially advanced technique is not clear. In general, the solid shading type phenonema that advanced techniques are typically designed to handle is more likely to arise in the residential environment due to artefacts in that environment which cannot easily be removed, so this may be a more appropriate place for these advanced techniques to be applied.

CONCLUSION

This chapter has considered the differences in large-scale (utility) and small-scale (residential) PV systems and what this suggest for the optimal choice of MPPT in each system. Key results referenced in the paper have shown that shadow from static objects in the environment causes the most severe non-uniform P-V characteristics for a PV module leading to the recommendation that in environments, typically residential, where there is likely to be sustained shading from static objects throughout the day advanced techniques may be more appropriate. From a literature survey, it was seen that most papers referring to a grid integrated, or utility size system typically focused on the connection of the system into the grid and considerations such as ramp rate control rather than on the specific MPPT method applied. As cloud transients are only likely to affect a small portion of the day and the need to curtail power due to ramp rate restrictions is a key consideration in utility scale PV systems, the use of conventional MPPT methods appears appropriate in these environments.

The key outcome of this chapter is showing that MPPT methods can be scalable to larger systems, however without other considerations such as the configuration of the system or the use of DMPPT this will not necessarily lead to better outcomes in the system.

REFERENCES

[1] Lyden S. and M. E. Haque, "Maximum Power Point Tracking techniques for photovoltaic systems: A comprehensive review and comparative analysis," *Renew. Sustain. Energy Rev.*, vol. 52, pp. 1504–1518, Dec. 2015.

[2] Subudhi B. and R. Pradhan, "A comparative study on maximum power point tracking techniques for photovoltaic power systems," *IEEE Trans. Sustain. Energy*, vol. 4, no. 1, pp. 89–98, Jan. 2013.

[3] Esram T. and P. L. Chapman, "Comparison of photovoltaic array maximum power point tracking techniques," *IEEE Trans. Energy Convers.*, vol. 22, no. 2, pp. 439–449, 2007.

[4] SunPower, *Highest efficiency Maxeon solar panels* | SunPower UK, 2020. [Online]. Available: https://www.sunpower.com.au/solar-panel-products/sunpower-maxeon-solar-panels. [Accessed: 08-Apr-2020].

[5] LG, *LG355N1C - Low Cost & Efficient Solar Panels* | *LG Solar Energy Australia*. [Online]. Available: https://www.lgenergy.com.

au/products/solar-panels/lg-neon-r-2-residential-60-cell/lg355n1c-1. [Accessed: 08-Apr-2020].

[6] NREL, *Best Research-Cell Efficiency Chart | Photovoltaic Research*. [Online]. Available: https://www.nrel.gov/pv/cell-efficiency.html. [Accessed: 08-Apr-2020].

[7] Sánchez-Carbajal S. and P. M. Rodrigo, *Optimum Array Spacing in Grid-Connected Photovoltaic Systems considering Technical and Economic Factors*, 2019.

[8] Olalla, C., C. Deline, D. Clement, Y. Levron, M. Rodriguez, and D. Maksimovic, "Performance of Power-Limited Differential Power Processing Architectures in Mismatched PV Systems," *IEEE Trans. Power Electron.*, vol. 30, no. 2, pp. 618–631, Feb. 2015.

[9] Sharma P. and V. Agarwal, "Exact Maximum Power Point Tracking of Grid-Connected Partially Shaded PV Source Using Current Compensation Concept," *IEEE Trans. Power Electron.*, vol. 29, no. 9, pp. 4684–4692, Sep. 2014.

[10] MacAlpine, S. M., R. W. Erickson, and M. J. Brandemuehl, "Characterization of Power Optimizer Potential to Increase Energy Capture in Photovoltaic Systems Operating Under Nonuniform Conditions," *IEEE Trans. Power Electron.*, vol. 28, no. 6, pp. 2936–2945, Jun. 2013.

[11] Hanson, A. J., C. A. Deline, S. M. MacAlpine, J. T. Stauth, and C. R. Sullivan, "Partial-Shading Assessment of Photovoltaic Installations via Module-Level Monitoring," *IEEE J. Photovoltaics*, vol. 4, no. 6, pp. 1618–1624, Nov. 2014.

[12] Dadkhah J. and M. Niroomand, "Optimization of MPPT in Both Steady and Transient States Through Variable Parameters," in *26th Iranian Conference on Electrical Engineering, ICEE 2018*, 2018, pp. 1132–1137.

[13] Arab N., B. Kedjar, A. Javadi, and K. Al-Haddad, "A Multifunctional Single-Phase Grid-Integrated Residential Solar PV Systems Based on LQR Control," *IEEE Trans. Ind. Appl.*, vol. 55, no. 2, pp. 2099–2109, Mar. 2019.

[14] Pandey A., N. Dasgupta, and A. K. Mukerjee, "High-Performance Algorithms for Drift Avoidance and Fast Tracking in Solar MPPT System," *IEEE Trans. Energy Convers.*, vol. 23, no. 2, pp. 681–689, Jun. 2008.

[15] Mujumdar U. B., and D. R. Tutkane, "Parallel MPPT for PV based residential DC Nanogrid," in *2015 International Conference on Industrial Instrumentation and Control, ICIC 2015*, 2015, pp. 1350–1355.

[16] Raiker G. A., L. Umanand, and B. Subba Reddy, "Perturb and Observe with Momentum Term applied to Current Referenced Boost Converter for PV Interface," in *Proceedings of 2018 IEEE International Conference on Power Electronics, Drives and Energy Systems, PEDES 2018*, 2018.

[17] Kollimalla S. K., and M. K. Mishra, "Variable Perturbation Size Adaptive P&O MPPT Algorithm for Sudden Changes in Irradiance," *IEEE Trans. Sustain. Energy*, vol. PP, no. 99, pp. 1–1, 2014.

[18] Ahmed E. M., and M. Shoyama, "Stability study of variable step size incremental conductance/impedance MPPT for PV systems," in *8th International Conference on Power Electronics - ECCE Asia: "Green World with Power Electronics," ICPE 2011-ECCE Asia*, 2011, pp. 386–392.

[19] Wang F., R. Gou, T. Zhu, Y. Yang, and F. Zhuo, "Comparison of DMPPT PV system with different topologies," in *China International Conference on Electricity Distribution, CICED*, 2016, vol. 2016-September.

[20] Zapata J. W., T. A. Meynard, and S. Kouro, "Partial power DC-DC converter for large-scale photovoltaic systems," in *2016 IEEE 2nd Annual Southern Power Electronics Conference, SPEC 2016*, 2016.

[21] Caiza D. L., S. Kouro, F. Flores-Bahamonde, and R. Hernandez, "Unfolding PV Microinverter Current Control: Rectified Sinusoidal vs Sinusoidal Reference Waveform," in *2018 IEEE Energy Conversion Congress and Exposition, ECCE 2018*, 2018, pp. 7094–7100.

[22] Rallabandi V., O. M. Akeyo, N. Jewell, and D. M. Ionel, "Incorporating battery energy storage systems into multi-MW grid connected PV systems," in *IEEE Transactions on Industry Applications*, 2019, vol. 55, no. 1, pp. 638–647.

[23] Chen J., D. Jiang, R. Yin, and Y. Liang, "New modelling of photovoltaic power integration system and its control strategy for medium-voltage DC distribution network," *J. Eng.*, vol. 2019, no. 17, pp. 3912–3917, Jun. 2019.

[24] Singh B. and P. Shukl, "Control of Grid Fed PV Generation using Infinite Impulse Response Peak Filter in Distribution Network," *IEEE Trans. Ind. Appl.*, pp. 1–1, Jan. 2020.

[25] Akeyo O. M., V. Rallabandi, N. Jewell, and D. M. Ionel, "The Design and Analysis of Large Solar PV Farm Configurations with DC Connected Battery Systems," *IEEE Trans. Ind. Appl.*, pp. 1–1, Jan. 2020.

[26] Craciun B. I., S. Spataru, T. Kerekes, D. Sera, and R. Teodorescu, "Power ramp limitation and frequency support in large scale PVPPs without storage," in *Conference Record of the IEEE Photovoltaic Specialists Conference*, 2013, pp. 2354–2359.

[27] Chen X., Y. Du, and H. Wen, "Forecasting based power ramp-rate control for PV systems without energy storage," in *2017 IEEE 3rd International Future Energy Electronics Conference and ECCE Asia, IFEEC - ECCE Asia 2017*, 2017, pp. 733–738.

[28] Perdana Y. S., S. M. Muyeen, A. Al-Durra, H. K. Morales-Paredes, and M. G. Simoes, "Direct connection of supercapacitor-battery hybrid storage system to the grid-tied photovoltaic system," *IEEE Trans. Sustain. Energy*, vol. 10, no. 3, pp. 1370–1379, Jul. 2019.

[29] Elgendy M. A., B. Zahawi, and D. J. Atkinson, "Assessment of the Incremental Conductance Maximum Power Point Tracking Algorithm," *IEEE Trans. Sustain. Energy*, vol. 4, no. 1, pp. 108–117, Jan. 2013.

[30] Lopez-Lapeña O. and M. T. Penella, "Low-power FOCV MPPT controller with automatic adjustment of the sample&hold," *Electron. Lett.*, vol. 48, no. 20, p. 1301, 2012.

[31] Ishaque K. and Z. Salam, "A deterministic particle swarm optimization maximum power point tracker for photovoltaic system under partial shading condition," *IEEE Trans. Ind. Electron.*, vol. 60, no. 8, pp. 3195–3206, 2013.

[32] Lyden S. and M. E. Haque, "A Simulated Annealing Global Maximum Power Point Tracking Approach for PV Modules under Partial Shading Conditions," *IEEE Trans. Power Electron.*, vol. 31, no. 6, 2016.

[33] Femia N., G. Lisi, G. Petrone, G. Spagnuolo, and M. Vitelli, "Distributed maximum power point tracking of photovoltaic arrays: novel approach and system analysis," *IEEE Trans. Ind. Electron.*, vol. 55, no. 7, pp. 2610–2621, 2008.

[34] Belhachat F. and C. Larbes, "Modeling, analysis and comparison of solar photovoltaic array configurations under partial shading conditions," *Sol. Energy*, vol. 120, pp. 399–418, 2015.

[35] Elgendy M. A., B. Zahawi, and D. J. Atkinson, "Assessment of perturb and observe MPPT algorithm implementation techniques for PV pumping applications," *IEEE Trans. Sustain. Energy*, vol. 3, no. 1, pp. 21–33, Jan. 2012.

[36] Manganiello P., M. Ricco, G. Petrone, E. Monmasson, and G. Spagnuolo, "Optimization of Perturbative PV MPPT Methods Through On Line System Identification," *IEEE Trans. Ind. Electron.*, vol. PP, no. 99, pp. 1–1, 2014.

[37] Lyden S. and M. E. Haque, "A hybrid simulated annealing and perturb and observe method for maximum power point tracking in PV systems under partial shading conditions," in *2015 Australasian Universities Power Engineering Conference: Challenges for Future Grids, AUPEC 2015*, 2015.

[38] Yi-Hwa L., H. Shyh-Ching, H. Jia-Wei, and L. Wen-Cheng, "A particle swarm optimization-based maximum power point tracking algorithm for PV systems operating under partially shaded conditions," *IEEE Trans. Energy Convers.*, vol. 27, no. 4, pp. 1027–1035, 2012.

[39] Mohanty S., B. Subudhi, and P. K. Ray, "A new MPPT design using grey Wolf optimization technique for photovoltaic system under partial shading conditions," *IEEE Trans. Sustain. Energy*, vol. 7, no. 1, pp. 181–188, Jan. 2016.

[40] Sundareswaran K., S. Peddapati, and S. Palani, "MPPT of PV Systems Under Partial Shaded Conditions Through a Colony of Flashing Fireflies," *IEEE Trans. Energy Convers.*, vol. 29, no. 2, pp. 463–472, Jun. 2014.

[41] Abdelrahman H., F. Berkenkamp, J. Poland, and A. Krause, "Bayesian optimization for maximum power point tracking in photovoltaic power plants," in *2016 European Control Conference (ECC)*, 2016, pp. 2078–2083.

[42] Lyden S., W. Olding, and Z. Darbari, "Bayesian Optimization for Maximum Power Point Tracking in Photovoltaic Systems," in *IEEE Power and Energy Society General Meeting*, 2018, vol. 2018-Augus.

[43] Galligan H. and S. Lyden, "Improving the Performance of Time Invariant Maximum Power Point Tracking Methods," in *2017 Australasian Universities Power Engineering Conference, AUPEC 2017*, 2017, pp. 1–6.

[44] Lyden S. and M. E. Haque, "Comparison of the perturb and observe and simulated annealing approaches for maximum power point tracking in a photovoltaic system under Partial Shading Conditions," in *2014 IEEE Energy Conversion Congress and Exposition, ECCE 2014*, 2014.

[45] Shenoy P. S., K. A. Kim, B. B. Johnson, and P. T. Krein, "Differential power processing for increased energy production and reliability of photovoltaic systems," *IEEE Trans. Power Electron.*, vol. 28, no. 6, pp. 2968–2979, 2013.

[46] Jazayeri M., S. Uysal, and K. Jazayeri, "A comparative study on different photovoltaic array topologies under partial shading conditions," in *2014 IEEE PES T&D Conference and Exposition*, 2014, pp. 1–5.

[47] Lappalainen K. and S. Valkealahti, "Effects of PV array layout, electrical configuration and geographic orientation on mismatch

losses caused by moving clouds," *Sol. Energy*, vol. 144, pp. 548–555, Mar. 2017.

[48] Chen X., Y. Du, H. Wen, L. Jiang, and W. Xiao, "Forecasting-based power ramp-rate control strategies for utility-scale PV systems," *IEEE Trans. Ind. Electron.*, vol. 66, no. 3, pp. 1862–1871, Mar. 2019.

[49] Lyden S. and M. E. Haque, "A hybrid simulated annealing and perturb and observe method for maximum power point tracking in PV systems under partial shading conditions," in *Australasian Universities Power Engineering Conference (AUPEC)*, 2015, pp. 1–6.

[50] Lappalainen K. and S. Valkealahti, "Recognition and modelling of irradiance transitions caused by moving clouds," *Sol. Energy*, vol. 112, pp. 55–67, Feb. 2015.

[51] Lappalainen K. and S. Valkealahti, "Apparent velocity of shadow edges caused by moving clouds," *Sol. Energy*, vol. 138, pp. 47–52, Nov. 2016.

[52] Lappalainen K. and S. Valkealahti, "Analysis of shading periods caused by moving clouds," *Sol. Energy*, vol. 135, pp. 188–196, Oct. 2016.

[53] Lyden S. and M. E. Haque, "Modelling, parameter estimation and assessment of partial shading conditions of photovoltaic modules," *J. Mod. Power Syst. Clean Energy*, Oct. 2018.

[54] Elasser A., M. Agamy, J. Sabate, R. Steigerwald, R. Fisher, and M. Harfman-Todorovic, "A comparative study of central and distributed MPPT architectures for megawatt utility and large scale commercial photovoltaic plants," in *IEEE Industrial Electronics Society Conference (IECON)*, 2010, pp. 2753–2758.

[55] Poshtkouhi S., V. Palaniappan, M. Fard, and O. Trescases, "A general approach for quantifying the benefit of distributed power electronics for fine grained MPPT in photovoltaic applications using 3D modeling," *IEEE Trans. Power Electron.*, vol. 27, no. 11, pp. 4656–4666, 2012.

In: Maximum Power Point Tracking
Editor: Maurice Hébert
ISBN: 978-1-53618-164-7
© 2020 Nova Science Publishers, Inc.

Chapter 5

IMAGE BASED MAXIMUM POWER POINT TRACKING IN WIND ENERGY CONVERSION SYSTEMS

K. Sujatha[1,*], M. Malathi[2], N. P. G. Bhavani[3,4] and V. Srividhya[4]

[1]EEE Dept, MGR Educational and Research Institute,
Chennai, Tamil Nadu, India
[2]ECE Dept, Chennai Institute of Technology, Chennai,
Tamil Nadu, India
[3]ECE Dept, MGR Educational and Research Institute,
Chennai, Tamil Nadu, India
[4]EEE Dept, Meenakshi College of Engineering,
Chennai, Tamil Nadu, India

---

[*] Corresponding Author's E-mail: drksujatha23@gmail.com.

Abstract

This manuscript deals with a simplified control stratagem to offer optimal power output power from a variable speed grid connected Wind Energy Conversion System (WECS).

A permanent magnet synchronous generator (PMSG) with variable speed turbine is coupled to the gear box, a bridge wave rectifier with blocking diode, a dc-to-dc converter with current controlled voltage source inverter which is of boost type.

Output power can be maximized using Image based Maximum Power Point Tracking (IMPPT) algorithm from the wind turbine operating in the range of cut-in to rated wind velocity which is sensed using Discrete Fourier Transform (DFT). The IMPPT algorithm, DC–DC and DC–AC converters with Fuzzy Logic Control (FLC) are simulated using MATLAB software. The obtained simulation results show that the identification of potential windmills is very important with respect to grid integration.

Keywords: Image processing, Fuzzy Logic Control, optimal wind energy

Introduction

Growing unconstructive effects of fossil fuel burning on the surroundings in addition to its restricted reserve have enforced several countries to search and revolutionize environmentally open alternatives that are renewable to prolong the increasing energy crisis. Changing to replaceable sources and the implementation of effective conservation measures would ensure sustainability.

Alternatives to conventional energy sources, particularly the renewable ones, are becoming progressively more attractive because of the limited fossil fuel reserves and the adverse effects associated with their use. The renewable energy resources include solar, wind, wave, geothermal and bio-energy. All these renewable energy resources are abundantly available in our country.

If these resources are well established, they can provide complete security of energy supply.

IMPORTANCE OF WIND ENERGY BASED ON SURVEY

At present, the wind energy is one of the best ever budding renewable energy source technologies across the world. Wind energy is a substitute for naturally replenished source compared to fossil fuel, which contaminates the lower layer of the atmosphere.

It has the benefit of being appropriate to be used in the vicinity of rural and remote areas. The rising requirement for energy supply joined with partial energy resources creates an exigency to discover novel solutions for this energy scarcity.

These days, wind energy analysis gives extraordinary information to researchers involved in renewable energy studies. The use of wind energy can considerably decrease the combustion of fossil fuel and the resultant emissions of various flue gases which carbon dioxide, carbon monoxide, sulphur dioxide and oxides of nitrogen.

In addition the usage of renewable sources of energy has become essential due to the present energy demand which is growing with environmental consciousness.

Knowledge of the statistical properties of the wind speed is necessary for forecasting the energy output from a Wind Energy Conversion System (WECS).

Because of the high inconsistency in space and time of wind energy, it is significant to verify that the analyzing method used for the measuring wind data will yield the estimated energy collected that is close to the actual energy collected. In recent years many efforts have been made to build abundant types of model for the wind speed frequency distribution [1-3].

The wind speed distribution, one of the wind characteristics, is of great importance for not only for structural and environmental design and analysis, but also for the assessment of the wind energy potential and the performance of wind energy conversion system as well.

Over the last two decades many findings have been devoted to develop an adequate statistical model to describe wind speed frequency distribution [4, 5].

RESEARCH HIGHLIGHTS

The literature survey indicates that a lot of efforts are taken to improve the control strategy and performance of the wind power generation system under various fluctuating conditions. The first objective deals with modern control system theory which has achieved much great importance, nowadays; because of its great impact in performance and thereby affecting the dynamic response of wind energy conversion systems during the last two decades such as optimal control methods.

The prediction of Wind speed estimation plays a major role in enhancing the wind power output. The change in orientation of the aerodynamic blades such that the incidents wind causes an airflow which is not same as the airflow far away from the turbine. This very nature states that the energy extracted also causes the same air to be deflected by the turbine. The centrifuged force in the rotating blades is proportional to the square of rotational speed of the blades which in turn increases the wind power as the cube of the wind speed. Hence the second objective of the research work would be to estimate the maximum output power corresponding to the wind velocity using Discrete Fourier Transform (DFT).

The last part of this research would be to offer an appropriate control strategy using Fuzzy logic for estimation optimal wind power speed. Also an Image Processing Based Identification of Potential windmills for grid integration is done here.

METHODS

The half an hour period wind speed data is used in this investigation and was monitored from the authorized weather stations installed. These weather stations were established in a suitable site such that the open space is free from obstacles like mountains, tall buildings, towers etc. Weather monitoring is carried out with a perspective to support joint research. The wind speed measurement was done using cup-anemometer mounted at a height of 10m above the ground along with other sensors was connected to a data logger. Every half an hour values of wind speed were then calculated and stored in sequence in the logger equipped with Random Access Memory (RAM). The data which was extracted was transferred to the laptop using RS-232 serial cable. The data tabulated was segregated normalized in to mean wind speed frequency and analyzed using FLC with Matlab software [6, 7]. The block diagram is shown in Figure 1(a) and (b) for capturing the images of Wind Turbines (WTs) with its related parameters and the corresponding values are tabulated in Table 1.

On the other hand the camera mounted on a clamped arrangement records the video of the moving WT blades. The corresponding angular frequency is measured using an accelerometer is attached to a rotating hub of a wind turbine measuring vibrations of the hub.

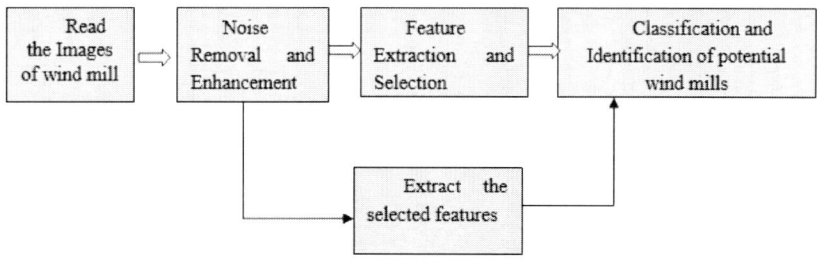

Figure 1(a). Methodology for Identification of Potential Windmills using Image Processing and FLC.

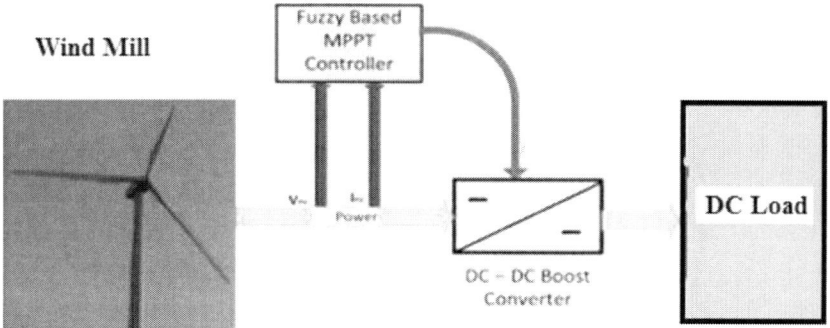

Figure 1(b). Image based Fuzzy Controller for MPPT in Windwills.

Table 1. Measurements pertaining wind mill Images captured

S. No	Images of Windmills	Angular frequency (rad/sec)	Blade sweep (m²)	Power Coefficient (no units)	Wind speed (m/s)
1.		2.38	0.60	0 to 0.2	1.7 to 1.9
2.		2.49	7.48	0.4 to 0.6	2 to 2.3
3.		2.76	10.68	0.7 to 1	2.4 to 2.7

To extract vibration signals originating from each blade, the vibration signals are combined with measuring signals from the azimuth angle sensor. Alternatively, the azimuth angle is found by using the measured accelerations in two directions representing the centrifugal forces and vibrating forces of the hub and feeding these signals into a phase locked loop unit establishing a phase of the rotating hub representing the azimuth angle. To extract each blade frequency, a Fast Fourier Transformation is used on the signals from the accelerometer [8, 9].

Blade Vision is claimed to minimize yield losses by near-eliminating yaw and pitch errors and pitch-angle asymmetries, and by (indirectly) indicating blade surface damage, icing build-up during blade rotation, and faults in turbine control parameters. Structural blade damage such as delamination is indicated by deviations in system output readings when the damage is sufficiently large to affect blade stiffness [10].

RESULTS AND DISCUSSION

A comparative analysis for Weibull Probability and AI based intelligent monitoring scheme for optimal wind power generation is carried out here to substantiate that this proposed method is an efficient method. This section includes the outcomes and the supporting arguments for the above mentioned two methods.

Results and Discussion for Weibull Probability Analysis

The Weibull distribution is the usually practiced distribution for modeling consistency statistics. This allocation is simple to construe and very flexible. The reliability studies offer the following information for optimal operation and identification of potential wind turbines. They are as follows

- The failure to produce optimal wind power output
- To propose optimal control scheme for continuous and online monitoring
- Identification of flaws in wind turbine

The Weibull distribution can model data that are right-skewed, left-skewed, or symmetric. Consequently, this allocation is used to assess reliability across varied applications, including vacuum tubes, capacitors, ball bearings, relays, and material strengths.

The Weibull distribution is also used to model a hazard function that is decreasing, increasing or constant, allowing it to describe the life span of a product or machinery.

The Weibull distribution may not work as effectively for product failures that are caused by chemical reactions or a degradation process like corrosion. Hence forth it is suggested for identification of potential windmills using image processing and AI based wind speed estimation.

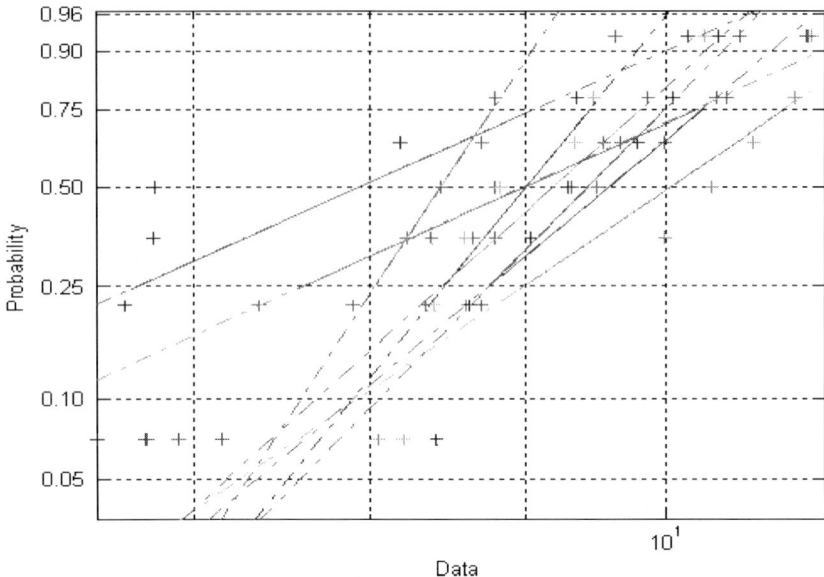

Figure 2. Weibull Probability Plot.

The distribution shows in Figure 2 that the probability of the predicted optimal wind power output is in the range of 0.5 to 0.96. Few samples were found to produce erroneous results which are scattered in the region of 0.05 to less than 0.5.

The total number of windmills monitored is 51 out of which 29 WTs are said to yield low power output.

Results and Discussion for Image Processing and FLC Based Identification

Nearly 51Wind Turbines (WT) were monitored to identify the potential wind mills.

The Figure 3 below illustrates that around 3 WTs have a power coefficient in the range of 0 to 0.2, 4 WTs have a power coefficient in the range of 0.4 to 0.6 and the remaining 44 WTs have a power coefficient in the range of 0.7 to 1 respectively.

Comparative Analysis of the Conventional and Proposed Scheme

The probability is calculated using Weibull distribution states that the failure for the WTs to yield maximized output is 0.568 and that of success rate is 0.432. This denotes that the correctly identified potential windmills are only 43.2% and the remaining 56.8% are found to be incorrectly classified. Similarly the probability is calculated using the proposed method states that the failure for the WTs to yield maximized output is 0.058 and that of success rate is 0.058. This denotes that the correctly identified potential windmills are only 94.1% and the remaining 5.88% are found to be incorrectly classified. These findings are illustrated in the Table 2.

The lowest value of the MSE of 0.099 is achieved around 50 iterations of training process. The Probability of the efficient classification is found to 0.9.

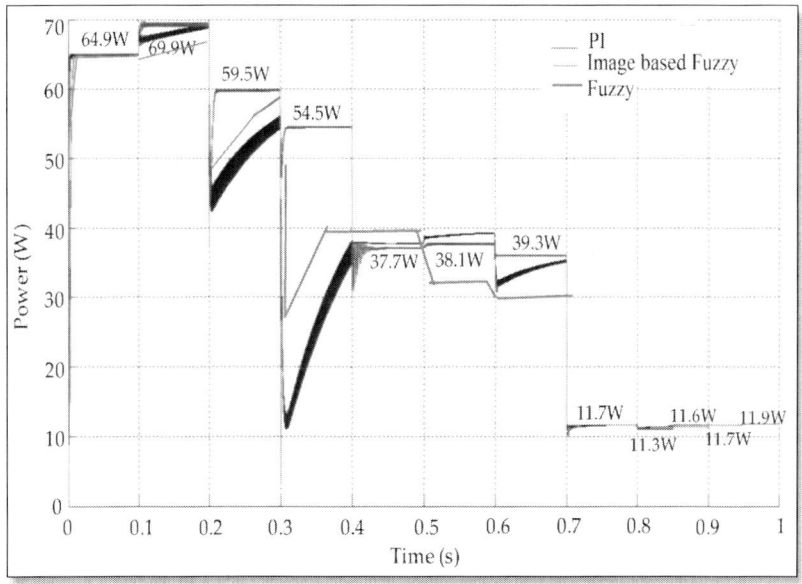

Figure 3. Estimation of maximum power output and Identification of Potential Windmills using Image Processing and FLC.

Table 2. Identification Efficiency

Weibull Probability Analysis			
Probability		Identification efficiency (%)	
Success	Failure	Correctly classified	Misclassified
0.432	0.568	43.2	56.8
Image Processing and FLC based Identification			
Success	Failure	Correctly classified	Misclassified
0.998	0.012	94.1	5.88
FLC based Identification			
Success	Failure	Correctly classified	Misclassified
0.898	0.212	91.2	8.8

CONCLUSION

Finally monitoring technique was employed for every critical location of a wind turbine. The common failures were overcome by using image processing and intelligent techniques. This technique would offer a feed forward control by identifying and proposing an optimal solution, to condition monitor with its intelligence control for the wind turbines for investigation of grid connected wind farms. Thus the main aim of this project is to determine the optimal site specific wind turbine design, which is the design that results in the lowest cost of energy at that particular site. There are many decisions that have to be made in designing a modern wind turbine. The optimal wind turbine design for one location is not necessarily the optimal design for another location because the wind speed distribution may vary between locations.

REFERENCES

[1] Weisser, D. A. (2003). Wind energy analysis of Grenada: an estimation using the 'Weibull' density function. *Renewable Energy*, 28: 1803 - 1812.

[2] Sathyajith, M., Pandey, K. P., Anil Kumar, V. (2002). Analysis of wind regimes for energy estimation. *Renewable Energy,* 25: 381 - 399.

[3] Bivona, S., Burlon, R., Leone Hourly (2003). Wind speed analysis in Sicily. *Renewable Energy,* 28: 1371 - 1385.

[4] Meishen, Li., Xianguo, Li. (2005). MEP-type distribution function: a better alternative to Weibull function for wind speed distributions. *Renewable Energy*, 30:1221 - 1240.

[5] Ramachandra, T. V., Shruthi, B. V. (2005). Wind energy potential mapping in Karnataka, India, using GIS. *Energy Conversion and Management,* 46: 1561 - 1578.

[6] Sujatha, K., Pappa, N., Combustion Quality Monitoring in PS Boilers Using Discriminant RBF, *ISA Transactions,* Vol. 2(7), (2011).

[7] Sujatha, K. and Dr. Pappa, N. Combustion Quality Estimation in Power Station Boilers using Median Threshold Clustering Algorithms, *International Journal of Engineering Science and Technology*, Vol. 2(7), 2623 - 2631, 2010.

[8] Bhavani, N. P. G., Dr. Sujatha, K. Soft Sensor for Temperature Measurement in Gas Turbine Power Plant, *International Journal of Applied Engineering Research*, pp. 21305 - 21316, 2014.

[9] Sujatha, K., Senthil Kumar, K., Godhavari, T. "An Effective Distributed Service Model for Image Based Combustion Quality Monitoring and Estimation in Power Station Boilers", *Lecture Notes in Electrical Engineering,* pp. 33 - 49, 2016.

[10] Gonzalez, R., López, L., Sánchez and Eduardo Calvo, "Angular Frequency Analysis of Wind Turbines by means of Digital Image Processing", *Recent Advances in Energy and Environment*, 229 - 233, 2012.

INDEX

A

algorithm, viii, x, 2, 8, 15, 16, 19, 22, 31, 35, 38, 40, 48, 51, 52, 57, 58, 59, 60, 70, 71, 96, 99, 100, 102, 105, 106, 107, 115, 144, 145, 149, 150, 159, 164
annealing, 145, 159, 160, 161
artificial intelligence, 70, 71

B

batteries, vii, viii, ix, 6, 7, 50, 65, 66, 67, 68, 81, 82, 83, 85, 86, 100, 102, 103, 110, 114, 115
battery life extension, 66
battery storage capacity, 115
biological processes, 142
body diode, 78, 123, 130

C

carbon dioxide, 165
carbon emissions, 2
carbon monoxide, 165
changing environment, 30, 40, 57, 147
chemical, 115, 169
chemical reactions, 115, 169
comparative analysis, 36, 48, 155, 169
complexity, 41, 48, 49, 138, 142, 144, 145, 148, 150
computing, 33, 34, 35, 36, 40, 42, 43, 48, 104
conditioning, 3, 10, 50, 85, 86, 99
conductance, 22, 24, 25, 57, 69, 105, 107, 144, 157
conduction, 4, 77, 78, 110, 123, 126, 130
conduction loss, 77, 78, 123, 126, 130
conference, 60, 103, 105
configuration, 10, 11, 12, 13, 93, 145, 146, 147, 149, 151, 155, 160
conversion efficiency, viii, ix, 2, 3, 72, 93, 94, 110, 116, 117, 123, 130, 131
converter, ix, x, 7, 9, 10, 11, 13, 25, 29, 30, 32, 37, 51, 52, 53, 58, 60, 66, 68, 69, 70, 71, 72, 73, 74, 75, 77, 78, 79, 93, 94, 95, 96, 98, 100, 102, 104, 106, 107, 110, 113, 118, 124, 126, 127, 128, 130, 142, 143, 145, 146, 147, 150, 157, 164
converter loss, 126, 127

cost, 3, 7, 10, 11, 13, 42, 43, 54, 97, 114, 138, 139, 148, 149, 150, 153, 172
cost saving, 3

D

DC-AC conversion efficiencies, 111, 119, 121, 122, 123, 126, 130
DC-AC converter, 6, 9, 10, 11, 13, 30, 111, 113, 124, 126, 127, 128, 129, 131
DC-DC efficiency, 126
degradation, 68, 126, 169
direct current-alternating current converter, vii, ix, 110
distribution, 36, 37, 53, 100, 145, 158, 165, 169, 170, 171, 172, 173
distribution function, 145, 173
drain, 76, 123

E

economic growth, 2, 49
efficiency of charging, 66
electricity, vii, viii, 3, 10, 50, 65, 66, 99, 114
electrolytic capacitor, 118
energy, viii, 2, 3, 6, 9, 10, 11, 12, 14, 39, 41, 44, 46, 49, 50, 53, 56, 103, 111, 114, 138, 139, 145, 146, 147, 148, 152, 153, 158, 160, 164, 165, 166, 172, 173
energy efficiency, 50
energy security, 2
energy supply, 164, 165
engineering, 105
environment, x, 137, 139, 141, 146, 147, 148, 151, 152, 154
environmental change, 23
environmental conditions, x, 14, 48, 137, 143, 144, 145, 147
environmental factors, 4
environmental impact, 138

environmental issues, 8
environments, 139, 148, 154
evolutionary computation, 60
extraction, 3, 27, 148, 150

F

field-effect transistor, vii, ix, 110, 111, 123
filter circuit, 118, 130
fluctuations, 53, 67, 69, 71, 78, 80
forward voltage, 73, 124, 125
fuzzy logic control, x, 33, 47, 56, 57, 70, 102, 105, 106, 164

H

health, vii, viii, ix, 66, 71
hybrid, 11, 12, 13, 50, 57, 67, 103, 107, 140, 146, 158, 159, 161

I

image, 116, 169, 172
image processing, 164, 166, 167, 170, 171, 172, 173
integration, x, 27, 150, 158, 164, 166
interface, 10, 117, 142, 152
inverter, ix, x, 11, 25, 27, 28, 54, 57, 62, 63, 100, 103, 110, 111, 112, 113, 117, 118, 123, 126, 130, 131, 164
irradiation, ix, 66, 67, 80, 85, 86, 88, 89, 90, 91, 92, 93, 100

L

lead-acid battery, 82, 83, 101, 102
light, 3, 32, 106, 116, 130
light-emitting diode, 116
lithium, vii, ix, 103, 110
lithium-ion batteries, vii, ix, 110

Index

load, viii, 1, 4, 7, 9, 10, 11, 12, 16, 26, 27, 29, 30, 45, 51, 62, 93, 94, 95, 96, 106, 114, 117, 118, 119, 121, 122, 123, 124, 126, 127, 130

M

metal-oxide-semiconductor, vii, ix, 110, 111
metal-oxide-semiconductor field-effect transistor (MOSFET), vii, ix, 75, 76, 77, 78, 98, 110, 111, 130, 134
methodology, 14, 20, 28, 39, 42, 66
modulation index, 124
modules, ix, 4, 5, 58, 60, 66, 79, 85, 86, 88, 100, 103, 139, 142, 146, 147, 150, 151, 161
MPPT algorithm, 8, 15, 16, 51, 100, 101, 102, 106, 107, 115, 157, 159
MPPT controller, 32, 114, 115, 116, 142, 158
multi-stage battery charging, 66

N

neural network, 35, 56, 57, 58, 70, 105
non-load loss, 118, 123, 124, 126, 127

O

on-resistance, ix, 77, 110, 113, 124, 125, 130, 131
optimal wind energy, 164
optimization, 29, 31, 35, 38, 40, 47, 48, 58, 59, 60, 80, 102, 104, 105, 144, 159, 160
oscillation, 28, 54, 59
output power, viii, x, 25, 30, 90, 93, 113, 117, 121, 122, 123, 125, 126, 127, 128, 129, 130, 145, 164, 166

P

partial shading, viii, 1, 2, 14, 35, 39, 40, 41, 43, 48, 57, 58, 59, 60, 101, 102, 139, 142, 159, 160, 161
photovoltaic, v, vii, viii, x, 1, 3, 4, 7, 10, 21, 50, 51, 52, 53, 54, 55, 56, 57, 58, 59, 60, 62, 65, 66, 99, 100, 101, 102, 103, 104, 105, 106, 107, 109, 110, 111, 112, 113, 131, 137, 138, 155, 156, 157, 158, 159, 160, 161
photovoltaic device, 110
photovoltaic power generation system, v, 55, 109, 111, 112, 113, 131
portable power generation system, 130
power generation, 3, 49, 50, 55, 58, 80, 100, 104, 111, 112, 113, 130, 131, 151
power loss, ix, 71, 72, 73, 75, 110, 111, 126, 128, 129, 130, 131, 143
power loss analysis, ix, 110, 131
power loss element, 126, 128, 129
power storage system, vii, ix, 110, 131
probability, 38, 39, 170, 171
probability density function, 38
prototype, 71, 72, 73, 75, 97
pulse-width modulation, 114

R

radiation, 3, 4, 11, 114, 117, 120
ramp, 92, 140, 141, 147, 148, 152, 155, 158, 161
recovery, ix, 78, 110, 117, 124, 125, 127, 130, 131
recovery loss, ix, 110, 117, 124, 127, 130, 131
redundancy, 146, 147, 152
reliability, 10, 110, 147, 152, 160, 169
renewable energy, 2, 11, 49, 50, 100, 138, 147, 164, 165
renewable energy technologies, 49, 50

requirements, x, 2, 110, 138, 139, 140, 141, 147, 148
resistance, ix, 77, 104, 110, 113, 124, 125, 130, 131
reverse recovery loss, ix, 110, 117, 124, 127, 130
reverse recovery time, 124, 125

S

scale system, x, 138, 140, 141, 142, 146, 147, 151, 152, 153
self-consumption, 96, 116
self-heating effect, 126
shading mitigation, vii, viii, 2, 62
si-based inverter, v, ix, 109, 110, 117, 123, 125, 126, 127, 131
SiC, v, vii, ix, 109, 110, 111, 112, 113, 117, 118, 119, 121, 122, 123, 124, 125, 126, 127, 128, 129, 130, 131, 134, 135
SiC inverter, 113, 117, 118, 126, 130
SiC MOSFET, ix, 110, 112, 113, 117, 125, 126, 130, 131
SiC-FET inverter, 112, 113
silicon carbide, 110
simulation, x, 39, 56, 164
small-scale, vii, x, 138, 140, 146, 153, 154
software, ix, x, 66, 85, 86, 87, 164, 167
solar cell, vii, ix, 3, 4, 6, 8, 9, 31, 32, 33, 86, 101, 105, 110, 111, 112, 113, 114, 130, 131, 132, 133
solar radiation power, 114, 117, 120
solution, 3, 22, 28, 31, 33, 37, 38, 39, 48, 84, 99, 153, 172
source, x, 7, 9, 11, 13, 41, 76, 77, 85, 111, 113, 123, 139, 156, 164, 165
specifications, 113, 147
spherical Si solar cell, vii, ix, 110, 112, 113, 130, 131
SPV array, viii, 2
steady state condition, 122

steady state measurement, 123
step load, 122, 123
storage, vii, viii, ix, 10, 11, 12, 18, 49, 50, 65, 66, 100, 110, 111, 115, 131, 148, 158
stratification, 69, 72, 84, 99
sustainability, 49, 164
sustainable energy, 50
switching frequency, 76, 110, 124, 130
switching loss, 106, 110, 124, 130
synchronous converter, 66
system analysis, 159

T

target, ix, 3, 66, 67, 70, 71, 78, 90, 100
techniques, vii, viii, ix, 2, 13, 15, 16, 36, 37, 39, 40, 42, 43, 48, 53, 54, 59, 62, 67, 69, 70, 71, 79, 85, 99, 100, 101, 104, 137, 138, 139, 140, 141, 142, 143, 144, 145, 146, 148, 149, 150, 151, 152, 154, 155, 159, 172
temperature, viii, ix, 1, 7, 8, 14, 16, 18, 21, 22, 33, 48, 56, 66, 68, 71, 75, 84, 86, 88, 94, 96, 100, 102, 105, 114, 116, 117, 120
temperature and humidity, 117, 120
terminals, 7, 72, 118, 119
thermal resistance, 126
thermal stability, 110
topology, 62, 140, 142, 145, 146, 148, 153, 154
turn-off loss, 124, 125, 126
turn-on loss, 124, 125, 126

U

utility-scale, 138, 161

V

velocity, x, 36, 161, 164, 166

W

wind farm, 172
wind power, 166, 169, 170
wind speed, 165, 166, 169, 172, 173
wind turbines, 50, 138, 169, 172